高等职业教育食品类专业新形态教材

制糖专业英语

Special English for Sugar Engineering

主　编　黄　凯　莫　蓓　董　毅
副主编　宁方尧　杭方学　刘家粤　范艳华
　　　　陆慧娟　韦　静　龙咏航
主　审　谭冠晖　黄　洁

北京理工大学出版社
BEIJING INSTITUTE OF TECHNOLOGY PRESS

内 容 提 要

本书以制糖工艺流程及制糖生产技术为主线，围绕甘蔗制糖工艺的核心单元，重点介绍国内外制糖技术与装备相关知识、专业词汇的使用，以及训练学习者对制糖相关主题内容的简要口头输出能力。全书包括 8 个单元，分别是 Sugarcane、Sugar Making Brief、Extraction in Sugarcane Mill、Clarification、Evaporation、Crystallization、Centrifuging、Storage。每个单元由小节（Section）及文化背景介绍（Culture Background）构成，每个小节学习内容为对话、小短文，以及词汇表和对应练习题（Task 1 ～ 4）。

本书提供了浅显易懂且篇幅较多的对话，以及大量的练习，同时配套在线课程供学习者使用。本书可供制糖技术人员参考使用。

图书在版编目（CIP）数据

制糖专业英语 / 黄凯，莫蓓，董毅主编 . -- 北京：北京理工大学出版社，2024.5（2024.10 重印）
ISBN 978-7-5763-3364-0

Ⅰ . ①制… Ⅱ . ①黄… ②莫… ③董… Ⅲ . ①制糖－英语 Ⅳ . ① TS244

中国国家版本馆 CIP 数据核字（2024）第 032359 号

责任编辑：阎少华　　**文案编辑**：阎少华
责任校对：周瑞红　　**责任印制**：王美丽

出版发行 / 北京理工大学出版社有限责任公司
社　　址 / 北京市丰台区四合庄路 6 号
邮　　编 / 100070
电　　话 /（010）68914026（教材售后服务热线）
（010）63726648（课件资源服务热线）
网　　址 / http：//www.bitpress.com.cn

版 印 次 / 2024 年 10 月第 1 版第 2 次印刷
印　　刷 / 河北鑫彩博图印刷有限公司
开　　本 / 787 mm×1092 mm　1/16
印　　张 / 9.5
字　　数 / 210 千字
定　　价 / 42.00 元

前　言
PREFACE

甘蔗是一种可再生资源，其种植对环境友好。甘蔗制糖产业可以采用可持续的农业种植方式减少对土地和水资源的消耗，并注重环境保护和生态平衡。这十分符合党的二十大提出的“推动绿色发展，促进人与自然和谐共生”的观点。甘蔗制糖是我国对“一带一路”国家技术输出的成熟技术之一，我国制糖技术人员帮助亚非拉国家创造“甜蜜的事业”做出了扎实的贡献，这也符合党的二十大报告提出的“促进世界和平与发展，推动构建人类命运共同体”的观点。学好制糖专业英语，有利于制糖技术的交流与发展。

制糖专业英语是制糖相关专业学生在完成基础英语和专业课程学习后开设的一门课程，本课程开设的目的是提高学生阅读专业书籍、资料和文献及简要口头输出能力。通过对课程的学习，学生可以进一步提高阅读、理解和翻译能力，同时扩大知识面、开阔视野；增大词汇量、了解专业词汇的构成方法（构词法）；提高阅读理解能力、掌握一些特殊的专业知识表达方式，特别是在中英文表达时的主要差别；掌握科技英语特点，初步具备翻译专业方面、资料的能力和简单的书面表达能力。

本书在编写的时候结合教学条件，力求提高讲课效率，目的是让学生能够尽可能地掌握一定的阅读、理解制糖专业英语资料的能力。本书以制糖工艺流程及制糖生产技术为主线，重点让学生了解制糖工业发展的现状、专业词汇的使用，以及训练学生对制糖相关主题内容的简要口头输出能力。

全书由 8 个单元构成，分别是 Sugarcane、Sugar Making Brief、Extraction in Sugarcane

Mill、Clarification、Evaporation、Crystallization、Centrifuging、Storage。每个单元由小节（Section）及文化背景介绍（Culture Background）构成，每个小节学习内容为对话、小短文及词汇表和对应练习题（Task 1 ～ 4）。

在本书编写过程中，我们尽力确保内容的准确性、完整性及合理性，但难免存在一些疏漏之处，恳请各位读者批评指正。

编 者

目录 CONTENTS

Unit 1 Sugarcane ······ 1

Section 1 Discussing the differences between sugarcane, grass and bamboo ······ 1

Section 2 Discussing the differences between sugarcane and sugarbeet ······ 6

Section 3 Sugarcane ······ 10

Section 4 Identify the picture ······ 15

Unit 2 Sugar Making Brief ······ 18

Section 1 Process of extracting sugar from sugarcane ······ 18

Section 2 Steps in the making of raw cane sugar ······ 24

Unit 3 Extraction in Sugarcane Mill ······ 31

Section 1 Thinking about how many parts of a sugarcane mill and how do they work 31

Section 2 The working principle of a sugarcane mill ······ 36

Section 3 Talking about the working process of a mill workshop ······ 41

Section 4 Mill workshop ······ 45

Section 5 Juice extraction ······ 49

Section 6 Identify the picture ······ 56

Unit 4 Clarification ······ 59

Section 1 Discussing the process of clarifying sugarcane juice ······ 59

Section 2 Defecation ······ 64
Section 3 Clarification ······ 68
Section 4 Identify the picture ······ 79
Unit 5 Evaporation ······ **81**
Section 1 Discussing the principles of evaporation ······ 81
Section 2 Evaporation process ······ 86
Section 3 Identify the picture ······ 95
Unit 6 Crystallization ······ **98**
Section 1 Discussing the purpose and basic operation of the crystallization process ······ 98
Section 2 The boiling operation ······ 104
Section 3 Crystallizer ······ 112
Section 4 Crystallizers ······ 116
Section 5 Identify the picture ······ 121
Unit 7 Centrifuging ······ **126**
Section Centrifuging ······ 126
Unit 8 Storage ······ **134**
Section 1 Storage ······ 134
Section 2 Changes during storage ······ 139
参考文献 ······ **146**

Unit 1 Sugarcane

Objectives

After learning this unit, you will be able to

1. Expand your vocabulary about sugar crops;
2. Understand the differences among the sugar crops;
3. Summarize the main differences between sugarcane and the other plants.

Section 1 Discussing the differences between sugarcane, grass and bamboo

Characters:

Mr. Sweet—a sugar industry professional（制糖行业的专业人士）

Tang Qianxi—a student majoring in Sugar Engineering and being enthusiastic about sugar crops（糖业工程专业学生，对糖料作物有浓厚兴趣）

Tang Qianxi (Tang): Hi, Mr. Sweet! I have a question about **botany**[1]. Could you please enlighten（指导）me on the differences between **sugarcane**[2], grass and bamboo?

Mr. Sweet: Of course, Tang Qianxi! Although sugarcane, grass and bamboo may share some similarities in appearance, they actually **belong to**[3] different plant families and genera（种类）.

Tang: I see. Could you please explain the main differences between them?

Mr. Sweet: Certainly! Sugarcane belongs to the **Poaceae**[4] family and is a **perennial**[5], thick, grass-like plant that can grow up to several meters tall. On the other hand, grass and bamboo belongs to different families and they are herbaceous plants（草本植物）with at least a one-year **lifespan**[6].

Tang: Very interesting! Now, why is sugarcane considered suitable for making sugar?

Mr. Sweet: Sugarcane contains a high **concentration**[7] of **sucrose**[8]. The **stems**[9] of sugarcane **accumulate**[10] a **significant**[11] amount of sucrose. Additionally, the **fiber**[12] content in sugarcane stems is relatively low, making it easier to **extract**[13] and **process**[14] the sugar. Hence, sugarcane is widely **recognized as**[15] an ideal **raw material**[16]. for sugar production.

Tang: That's fascinating! I have gained a deeper understanding of sugarcane, grass and bamboo. Thank you for your help, Mr. Sweet!

Mr. Sweet: You're welcome, Tang Qianxi! I'm glad I could assist you. If you have any other questions, feel free to ask me anytime!

音频：Unit 1　Section 1

New words and expressions

1. botany	/'bɒtəni/	*n.* 植物学
2. sugarcane	/'ʃʊgəkeɪn/	*n.* [作物] 甘蔗
3. belong to		属于
4. Poaceae	/poʊ'eɪsiaɪ/	禾本科
5. perennial	/pə'reniəl/	*adj.* 多年生的
6. lifespan	/'laɪfspæn/	*n.* （人或动物的）寿命
7. concentration	/ˌkɒns(ə)n'treɪʃ(ə)n/	*n.* 浓度
8. sucrose	/'suːkrəʊz; 'suːkrəʊs/	*n.* 蔗糖
9. stem	/stem/	*n.*（植物的）茎
10. accumulate	/ə'kjuːmjəleɪt/	*v.* 积累
11. significant	/sɪg'nɪfɪkənt/	*adj.* 显著的
12. fiber	/'faɪbə(r)/	*n.* 纤维
13. extract	/'ekstrækt; ɪk'strækt/	*v.* 提取
14. process	/'prəʊses/	*v.* 加工
15. be recognized as		被认为是
16. raw material		原材料

Exercises

Task 1: Crosswords puzzle

请根据中文提示把下面字谜中的单词填写出来。

Across（横排）	Down（竖排）
6. 显著的	1. 加工
8. 甘蔗	2. 浓度
9. 积累	3. 多年生的
10. 提取	4. 蔗糖
11. 植物学	5. 茎
	7. 纤维

Task 2: One-choice question

根据对话内容，选择下列问题的正确答案。

1. Which of the following does NOT belong to the Poaceae family?________

A. Sugarcane. B. Grass.

C. Bamboo. D. Tomato.

2. What is the main reason why sugarcane is considered suitable for making sugar?________

A. It contains a high concentration of fructose. B. Its stems are easy to extract.

C. It belongs to the grass family. D. It is a perennial plant.

3. What is the scientific study of plants called?________

A. Botany.
B. Biology.
C. Chemistry.
D. Geology.

Task 3: Blank filling

根据对话内容，把下列句子补充完整，每空填写一个单词。

1. Sugarcane belongs to the________family.
2. Sugarcane accumulates a significant amount of________in its stems.
3. Sugarcane is widely recognized as an ideal raw material for________production.

Task 4: Thinking training

根据对话内容，绘制甘蔗与其他植物的区别的思维导图，并试着复述。

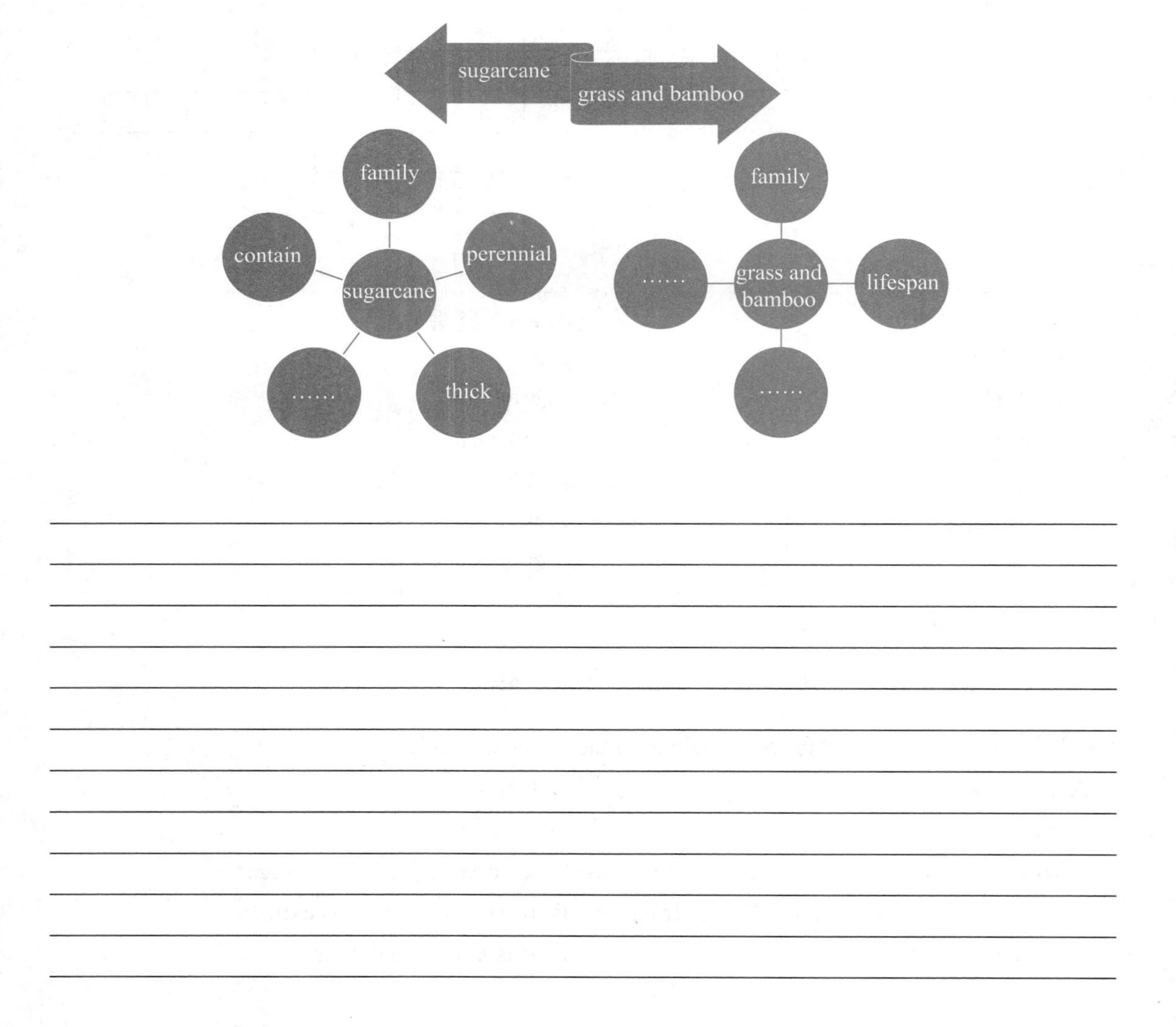

Key（Unit 1 Section 1）

Task 1:

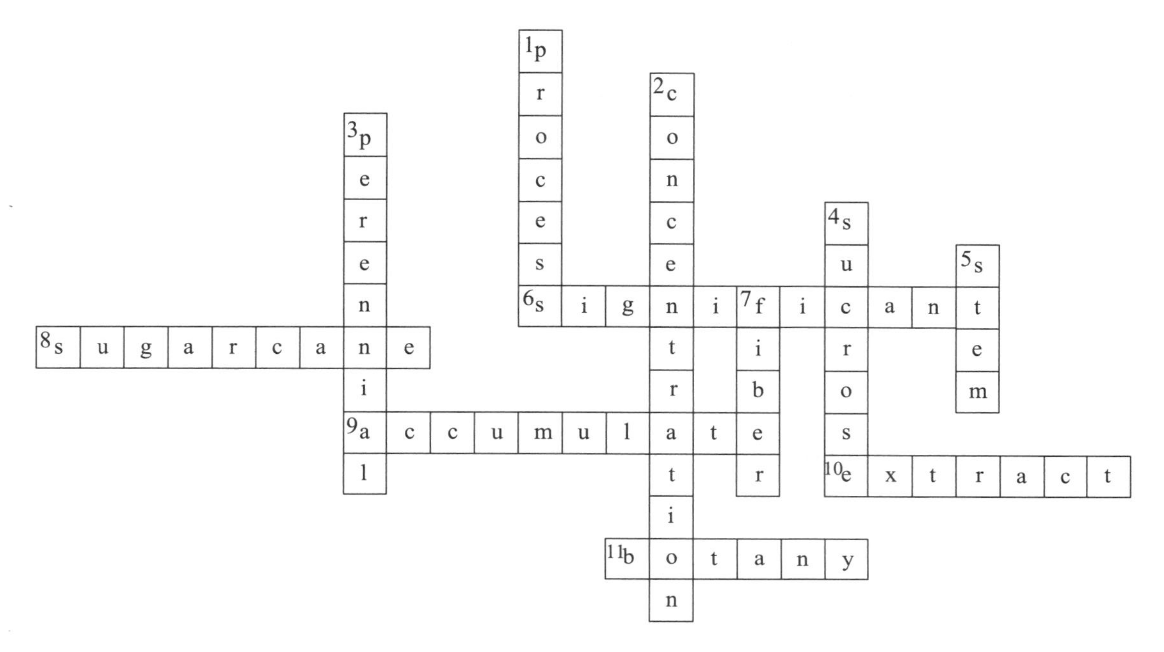

Task 2:

1. D 2. B 3. A

Task 3:

1. Poaceae 2. sucrose 3. sugar

Task 4:

The main differences between sugarcane and grass and bamboo are as follows: Firstly, they belong to different families and genera. Sugarcane is in the Poaceae family, while grass and bamboo are in different families. Secondly, sugarcane is a perennial thick grass-like plant that can grow several meters tall. Grass and bamboo are herbaceous plants with at least a one-year lifespan. Thirdly, sugarcane is suitable for making sugar due to its high sucrose content and relatively low fiber content in stem.

译文

第一节：讨论甘蔗、草和竹子的区别

唐千禧：你好，甜先生！我有一个关于植物学的问题。你能给我讲讲甘蔗、草和竹子的区别吗？

甜先生：当然可以了，唐千禧！虽然甘蔗、草和竹子在外观上可能有一些相似之处，但实际上它们属于不同的植物科和属。

唐千禧：我明白了。你能解释一下它们的主要区别吗？

甜先生：当然！甘蔗属于禾本科，是一种多年生的、厚的、像草一样的植物，可以长到几米高。另一方面，草和竹子属于不同的科，它们是草本植物，至少有一年的寿命。

唐千禧：真有意思！那为什么甘蔗被认为适合制糖呢？

甜先生：甘蔗含有高浓度的蔗糖。甘蔗的茎部积累了大量的蔗糖。此外，甘蔗茎中的纤维含量相对较低，更容易提取和加工成糖。因此，甘蔗被广泛认为是制糖的理想原材料。

唐千禧：太棒了！我对甘蔗、草、竹子有了更深的了解。谢谢你的帮助，甜先生！

甜先生：不客气，唐千禧！我很高兴能帮到你。如果你有任何其他问题，请随时问我！

Section 2 Discussing the differences between sugarcane and sugarbeet

Mr. Sweet: Good morning, Tang. How can I assist you today?

Tang Qianxi (Tang): Good morning, Mr. Sweet. I've always been fascinated by sugar crops. Can you please enlighten me on the differences between sugarcane and **sugarbeet**[1]?

Mr. Sweet: Certainly, sugarcane and sugarbeet are both used as sources of sugar, but they have distinct characteristics. One of the main differences is their origin and growth environments. Sugarcane is primarily grown in tropical and subtropical（热带及亚热带）regions, while sugar-beet **thrives**[2] in temperate climates（温和气候）with cooler summers.

Tang: Ah, interesting. Does their taste different as well?

Mr. Sweet: Indeed, it does. Sugarcane **is known for**[3] its naturally sweet taste and a unique hint of freshness. On the other hand, sugarbeet exhibits a milder sweetness with a slightly earthy or beet-like **flavor**[4]. The distinct taste of each crop often influences their usage in various **cuisines**[5].

Tang: That's fascinating. Are there any other differences between their textures（口感）or uses?

Mr. Sweet: Absolutely. Sugarcane is characterized by its **fibrous**[6] stalks(茎), **rich in**[7] sucrose content, which are typically crushed to extract the juice for sugar production. Sugarbeet, though not as fibrous, contains more sugar content in its root. It is usually processed to extract sugar **in the form of**[8] **granules**[9], largely used in food products and for **sweetening**[10] beverages（饮料，酒水）.

Tang: I see. So, their **cultivation**[11] regions also differ, right?

Mr. Sweet: Yes, indeed. As I mentioned earlier, sugarcane flourishes in tropical and subtropical regions, such as Southeast Asia, South America and parts of Africa. On the other hand, sugarbeet is commonly found in areas with temperate climates, like Europe, North America and Eastern Asia.

Tang: Thank you, sir. Your explanation has shed light on（阐明）the differences between sugarcane and sugarbeet. It's truly intriguing to learn about these fascinating plants that provide us with sweetness.

Mr. Sweet: You're welcome.

音频：Unit 1　Section 2

New words and expressions

1. sugarbeet	/ˈʃʊɡə biːt/	*n.* 甜菜
2. thrive	/θraɪv/	*v.* 茁壮成长
3. be known for		因……而众所周知
4. flavor	/ˈfleɪvə(r)/	*n.* 风味
5. cuisine	/kwɪˈziːn/	*n.* 烹饪
6. fibrous	/ˈfaɪbrəs/	*adj.* 纤维的
7. rich in		富含
8. in the form of		以…的形式
9. granule	/ˈɡrænjuːl/	*n.* 颗粒
10. sweeten	/ˈswiːtn/	*v.* 变甜
11. cultivation	/ˌkʌltɪˈveɪʃ(ə)n/	*n.* 种植

Exercises

Task1: Crosswords puzzle

请根据中文提示把下面字谜中的单词填写出来。

Across（横排）	Down（竖排）
5. 种植	1. 风味
6. 茁壮成长	2. 变甜
7. 颗粒	3. 烹饪风格
	4. 纤维的

Task 2: One-choice question

根据对话内容，选择下列问题的正确答案。

1. Sugarcane is primarily grown in ________.
   A. temperate climates
   B. tropical and subtropical regions
   C. Europe and North America
   D. Eastern Asia
2. Which crop has a slightly earthy or beet-like flavor? ________
   A. Sugarcane.
   B. Both sugar cane and sugarbeet.
   C. Neither sugarcane nor sugarbeet.
   D. Sugarbeet.
3. Sugarcane is known for its ________.
   A. fibrous stalks with rich sucrose content
   B. milder sweetness with an earthy flavor
   C. growth in temperate climates
   D. extraction of sugar in the form of granules
4. Sugarbeet is commonly found in areas with ________.
   A. tropical climate
   B. subtropical regions
   C. temperate climates
   D. Southeast Asia

Task 3: Blank filling

根据对话内容，把下列句子补充完整，每空填写一个单词。

1. Sugarcane is primarily grown in ________ and ________ regions.
2. Sugarbeet exhibits a milder sweetness with a slightly ________ flavor.
3. Sugarcane is characterized by its fibrous stalks, rich in ________ content.
4. Sugarbeet is usually processed to extract sugar in the form of ________.

Task 4: Reading comprehension

根据对话内容，回答下列问题。

1. According to the dialogue, where does sugarcane primarily flourish?

____

____

2. How does the taste of sugar beet differ from sugarcane?

____

____

Key (Unit 1　Section 2)

Task 1:

											1f				
			2s			3c					l		4f		
			w		5c	u	l	t	i	v	a	t	i	o	n
			e			l					v		b		
			e			s					o		r		
			6t	h	r	i	v	e			r		o		
			e			n							u		
7g	r	a	n	u	l	e							s		

Task 2:

1.B　2.D　3.A　4.C

Task 3:

1. tropical, subtropical
2. earthy or beet-like
3. sucrose
4. granules

Task 4:

1. In tropical and subtropical regions.
3. Sugarbeet has a milder sweetness with a slightly earthy or beet-like flavor.

译文

第二节：讨论甘蔗与甜菜的区别

甜先生：早上好，唐。今天有什么需要我帮助的吗？

唐千禧（唐）：早上好，甜先生。我一直对甜作物很着迷。您能给我解释一下甘蔗和甜菜的区别吗？

甜先生：当然可以。甘蔗和甜菜都是用来制糖的原料，但它们有明显的特点。其中一个主要的区别是它们的来源和生长环境。甘蔗主要生长在热带和亚热带地区，而甜菜则适宜在温和气候、夏天较凉爽的地区生长。

唐：啊，有趣。它们的口味也有所不同吗？

甜先生：确实如此。甘蔗以其天然甜味和独特的清新气息而闻名。而甜菜则带有较为温和的甜味，略带有土壤或甜菜根的风味。每种作物独特的口味经常会影响它们在各种烹饪中的用途。

唐：真有趣。它们的口感或用途还有其他区别吗？

甜先生：当然。甘蔗以纤维丰富的茎部为特征，富含蔗糖，通常被压榨以提取糖汁用于制糖。甜菜虽然不像甘蔗那样纤维丰富，但它的根部含有更多的糖分。它通常经过加工，以提取颗粒状糖的形式，广泛用于食品产品和饮料的甜味。

唐：我明白了。所以，它们的种植地区也有所不同，对吗？

甜先生：是的，正如我之前提到的，甘蔗在热带和亚热带地区茁壮成长，如东南亚、南美及非洲的部分地区。而甜菜通常在温和气候的地区，如欧洲、北美以及东亚等地比较常见。

唐：谢谢您，先生。您的解释阐明了甘蔗和甜菜的区别。学习这些为我们带来甜美的植物真令人着迷。

甜先生：不客气。

Section 3 Sugarcane

The sugarcane is a large perennial grass which **attains**[1] a length of 10 to 24 feet（英尺）at **maturity**[2]. There are a number of **species**[3] and **numerous**[4] **varieties**[5]. Sugarcane is **propagated**[6] by **cuttings**[7], each cutting **consisting of**[8] **portions of**[9] the **cane plant**[10] having two or more **buds**[11]. The buds develop into cane plants on which several other **shoots**[12] arise from below the **soil level**[13] to form **a clump of**[14] stalks. From 12 to 20 months are required for plant crops(new plants)and about 12 months for **ratoons**[15], that is, cane stalks arising from stalks which have been previously（先前地）**reaped**[16]. Fields arc replanted（重新播种）after 2 to 5 reapings have been obtained（获得）from the original cuttings. The cane stalk is round,1 to 3 inches（英寸）in **diameter**[17], and is covered with a hard **rind**[18] which is light brown, or green, yellowish green, or purple in color, depending on variety, the stalk consists of **a series of**[19] **joints**[20] or **internodes**[21]**separated**[22] by nodes (not unlike bamboo). The rind and the nodes are of a **woody**[23]nature, while the internode is soft **pith**[24]. The internode contains the greater part of the juice. In harvesting, the canes are cut at ground level, and the leaves and tops of the stalks are removed. The sucrose content of the **ripened**[25] cane varies between 10 to 15 percent of the total weight; 13 percent is about average.

音频：Unit 1 Section 3

New words and expressions

1. attain	/ə'teɪn/	*v.* 收获
2. maturity	/mə'tʃʊərəti/	*n.* 成熟
3. species	/'spiːʃiːz/	*n.*（动植物的）种，物种；种类
4. numerous	/'njuːmərəs/	*adj.* 众多的，许多的
5. variety	/və'raɪəti/	*n.* 品种，变化
6. propagate	/'prɒpəgeɪt/	*v.* 繁殖
7. cut	/kʌt/	*v.* 插枝
8. consist of		由……组成
9. portion of…		……部分
10. cane plant		蔗株
11. bud	/bʌd/	*n.* 芽，花蕾
12. shoot	/ʃuːt/	*n.* 嫩芽
13. soil level		土壤
14. a clump of		一丛
15. ratoon	/rə'tuːn/	*n.* 截根苗（农作物，尤指甘蔗的）
16. reap	/riːp/	*v.* 收获；收割（庄稼等）
17. diameter	/daɪ'æmɪtə(r)/	*n.* 直径
18. rind	/raɪnd/	*n.* 壳；外皮
19. a series of		一系列
20. joint	/dʒɔɪnt/	*n.* 关节
21. internode	/'ɪntəˌnəʊd/	*n.* 节间
22. separate	/'seprət/	*v.* 分离，分开
23. woody	/'wʊdi/	*adj.* 木质的
24. pith	/pɪθ/	*n.* 木髓
25. ripened	/'raɪpənd/	*adj.* 成熟的

Exercises

Task 1: Crosswords puzzle

请根据中文提示把下面字谜中的单词填写出来。

Across（横排）	**Down（竖排）**
6. 直径	1. 木质的
9. 成熟	2. 许多的
10. 品种，变化	3. 插枝
11. 物种，种类	4. 芽
	5. 壳，外皮
	7. 收割
	8. 木髓

Task 2: One-choice question

根据文章内容，选择下列问题的正确答案。

1. The length of the sugarcane at maturity is ________.

A. 1 to 3 inches　　B. 2 to 5 reapings

C. 10 to 24 feet　　D. 12 to 20 months

2. Sugarcane is propagated by ________.

A. cuttings　　B. plantings

C. sowings　　D. hydroponics

3. What's color of the rind of the sugarcane?________.

A. Light brown, or green　　B. Yellowish green

C. Purple　　D. All above

Task 3: Blank filling

根据文章内容，把下列句子补充完整，每空填写一个单词。

1. The sugarcane is a large perennial ________ which attains a length of 10 to 24 feet at maturity.
2. The rind and the nodes are of a ________ nature, while the internode is soft pith.
3. The internode contains the greater part of the ________ .

Task 4: Thinking training

根据文章内容，把下列关于甘蔗特征的表格填写完整，并尝试与搭档一起描述甘蔗的基本特征。

Features	Description
Definition 定义	· The sugarcane is a large ________ grass.
Height of plant 植物的高度	· It can reach ________ in length at maturity.
The way of propagating 繁殖的方式	· It is propagated by ________.
Morphological characteristics of plants 植物的形态特点	· The cane stalk is ________ , ________ in diameter, and is covered with a ________ . · The color is light brown or ________ , yellowish green or ________ . · The stem consists of a series of ________ or ________ . · The rind and the nodes are of a ________ nature, while the internode is ________ pith.
The value and uses 价值和用途	· Mature sugarcane contains ________ percent sucrose by weight

Key (Unit 1 Section 3)

Task 1:

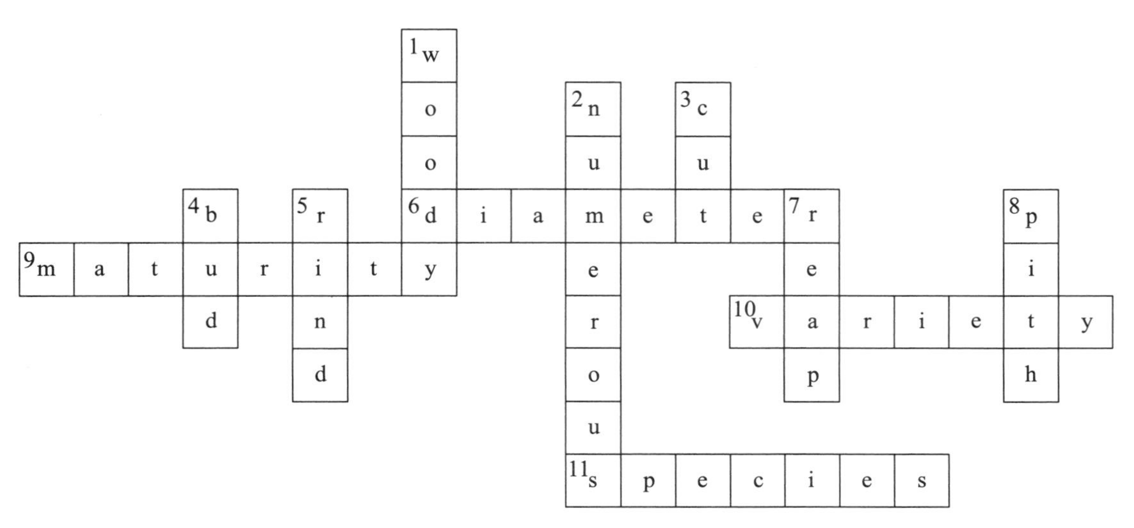

Task 2:

1. C 2. A 3. D

Task 3:

1. grass 2. woody 3. juice

Task 4:

Features	Description
Definition 定义	· The sugarcane is a large perennial grass.
Height of plant 植物的高度	· It can reach 10 to 24 feet in length at maturity
The way of propagating 繁殖的方式	· It is propagated by cuttings.
Morphological characteristics of plants 植物的形态特点	· The cane stalk is round, 1 to 3 inches in diameter, and is covered with a hard rind. · The color is light brown, or green, yellowish green, or purple. · The stem consists of a series of nodes or internodes. · The rind and the nodes are of a woody nature, while the internode is soft pith.
The value and uses 价值和用途	· Mature sugarcane contains 10 to 15 percent sucrose by weight.

译文

第三节：甘蔗

甘蔗是一种大型多年生草本植物，成熟收获时可达 10 ~ 24 英尺长。有许多种类和大量变种。甘蔗是通过插枝繁殖的，每次插枝由甘蔗植株的两个或两个以上的芽的一部分组成。这些芽发育成甘蔗植株，其他几个嫩芽从土壤下面长出来，形成一簇茎。作物（新植物）需要 12 ~ 20 个月，截根苗（即以前收获的甘蔗茎）需要大约 12 个月。原插枝收获 2 至 5 次后，进行重新播种。茎是圆形的，直径 1 ~ 3 英寸，并覆盖着一层坚硬的外皮，其颜色为浅棕色，或绿色，黄绿色，或紫色，视品种而定，茎由一系列节组成，或由节间分开（不像竹子）。皮和节为木质，节间为软木髓。节间含有大部分汁液。在收割时，甘蔗从地面高度被砍断，叶子和茎的顶部被移除。成熟甘蔗的蔗糖含量在总质量的 10% ~ 15% 之间，平均为 13% 左右。

Section 4 Identify the picture

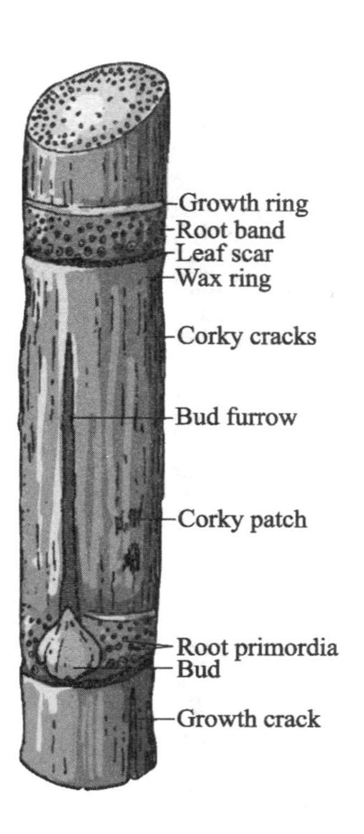

New words and expressions

1. bud /bʌd/ 芽孢
2. sugarcane stalk 甘蔗茎秆 (after Artschwager and Brandes 1958, from James 2004)
3. growth ring 生长轮
4. root band 根带
5. leaf scar 叶痕
6. 6.wax ring 蜡带
7. bud furrow 牙沟
8. corky patch 软木质斑块
9. root primordia 根源基
10. growth crack 生长裂纹

Exercises

Task1: Fill in the blanks

认真观察图片，并在空白处填上正确的单词

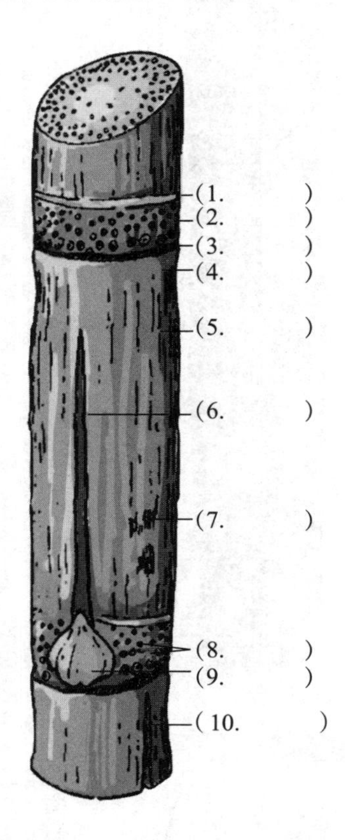

Key（Unit1 Section4）

Task1:

1. sugarcane stalk 甘蔗茎秆
2. growth ring 生长轮
3. root band 根带
4. leaf scar 叶痕
5. wax ring 蜡带
6. bud furrow 芽沟
7. corky patch 软木质斑块
8. root primordia 根源基
9. bud /bʌd/ 芽孢
10. growth crack 生长裂纹

Culture Background

Sugarcane Origin

Sugarcane probably originated in New Guinea or India, and later spread to the South Ocean islands. It was introduced to southern China around the time of King Xuan of the Zhou Dynasty.

The “zhe” in the pre-Qin period was sugarcane, and the word “sugarcane” did not appear until the Han Dynasty. The pronunciation of “zhe” and “sugarcane” may come from Sanskrit. From the 10th to the 13th century (Song Dynasty), sugarcane was widely cultivated in southern provinces. Indochina Peninsula and other places in South Asia such as Zhenla, Chamcheng, Sanfoqi, Sujidan are also widely planted sugarcane.

甘蔗起源

甘蔗原产地可能是新几内亚或印度，后来传播到南海群岛。大约在周宣王时传入中国南方。先秦时代的“柘”就是甘蔗，到了汉代才出现“蔗”字，“柘”和“蔗”的读音可能来自梵文。10 到 13 世纪（宋代），南方各省普遍种植甘蔗。中南半岛和南亚各地如真腊、占城、三佛齐、苏吉丹也普遍种甘蔗。

Unit 2 Sugar Making Brief

Objectives

After learning this unit, you will be able to

1. Expand your vocabulary about sugar making steps;
2. Summarize the process of extracting sugar form sugarcane:
3. Explain the steps of raw cane sugar making.

Section 1 Process of extracting sugar from sugarcane

Characters:
Mr. Sweet—a sugar industry professional （制糖行业的专业人士）
Tang Qianxi, Lin Huahua, Zhang Jiejing, Wang Tichun—four students majoring in Sugar Engineering and being enthusiastic about sugar crops (糖业工程专业学生，对糖料作物有浓厚兴趣)

Mr. Sweet: Good morning, class. Today, we are going to discuss the working principle of **sugar**[1] production. Let's start by understanding the process of extracting sugar from sugarcane.

Tang Qianxi: How does the extraction process work?

Mr. Sweet: It **begins with**[2] the sugarcane being crushed to extract the juice. The extraction is the crucial（至关重要的）step in obtaining（获取）the raw material for sugar production.

Lin Huahua: What happens next?

Mr. Sweet: After extraction, the juice **goes through**[3] a process called **clarification**[4]. It involves removing **impurities**[5] and **solid**[6] **particles**[7] from the juice, ensuring it becomes clear and clean.

Zhang Jiejing: What happens after clarification?

Mr. Sweet: The next step is **evaporation**[8]. In this process, the juice is heated to remove most of the water content, leaving behind a thick **syrup**[9] called **molasses**[10].

Wang Tichun: And how do we get sugar from the molasses?

Mr. Sweet: The molasses then undergoes **boiling**[11], where it is heated again to form sugar **crystals**[12]. This process is known as **crystallization**[13] or boiling the sugar.

Tang Qianxi: We have sugar crystals now. What's next?

Mr. Sweet: We need to **separate**[14] the sugar crystals from the remaining **liquid**[15]. This is done using a centrifuge, a machine that spins at high speed to separate the sugar crystals from the syrup. This step is called **centrifuging**[16] or separating the molasses from the sugar.

Lin Huahua: That's fascinating! So, we finally get pure sugar crystals?

Mr. Sweet: Almost there. The last step is **drying**[17] the sugar crystals to remove any remaining **moisture**[18]. Once dried, we have the **final product**[19] — sugar!

Zhang Jiejing: Wow! It's amazing to learn about the intricate（复杂的）**process**[20] behind sugar production.

Mr. Sweet: Indeed, it is a fascinating journey from sugarcane to the sweet crystals we all enjoy. Remember, understanding the process helps us appreciate the effort put into producing everyday items we often take for granted.

音频：Unit 2 Section 1

New words and expressions

1. sugar	/ˈʃʊɡə(r)/	*n.* 糖
2. begin with		以……开始；开始于……
3. go through		经过
4. clarification	/ˌklærəfɪˈkeɪʃn/	*n.* 澄清
5. impurity	/ɪmˈpjʊərəti/	*n.* 杂质
6. solid	/ˈsɒlɪd/	*adj.* 固态的
7. particle	/ˈpɑːtɪk(ə)l/	*n.* 颗粒
8. evaporation	/ɪˌvæpəˈreɪʃ(ə)n/	*n.* 蒸发
9. syrup	/ˈsɪrəp/	*n.* 糖浆
10. molasses	/məˈlæsɪz/	*n.* 糖蜜，糖浆

11. boiling	/ˈbɔɪlɪŋ/	*n.* 沸腾
12. crystal	/ˈkrɪst(ə)l/	*n.* 结晶，晶体
13. crystallization	/ˌkrɪstəlaɪˈzeɪʃ(ə)n/	*n.* 结晶化
14. separate	/ˈseprət/	*v.* 分离
15. liquid	/ˈlɪkwɪd/	*n.* 液体，液态物
16. centrifuging	/ˈsentrɪfjuːdʒ/	*v.* 离心法
17. dry	/draɪ/	*v.* 干燥
18. moisture	/ˈmɔɪstʃə(r)/	*n.* 潮气，水分
19. final product		最终产品
20. process	/ˈprəʊses/	*n.* 过程；加工方法

Exercises

Task 1: Crosswords puzzle

请根据中文提示把下面字谜中的单词填写出来。

Across（横排）	**Down（竖排）**
2. 澄清	1. 颗粒
5. 水分	3. 糖蜜
7. 分离	4. 沸腾
9. 过程	6. 蒸发
11. 结晶	8. 提取
13. 糖	10. 干燥
14. 杂质	12. 液体
	13. 固态的

Task 2: One-choice question

根据对话内容，选择下列问题的正确答案。

1. What is the process called when the juice is heated to form sugar crystals?________
   A. Clarification.
   B. Evaporation.
   C. Extraction.
   D. Crystallization.
2. What is the purpose of using a centrifuge in the sugar production process?________
   A. To extract juice from sugarcane.
   B. To clarify the sugarcane juice.
   C. To separate impurities from sugar crystals.
   D. To separate sugar crystals from molasses.
3. What is the final product obtained after drying the sugar crystals?________
   A. Syrup.
   B. Centrifuge.
   C. Raw material.
   D. Sugar.

Task 3: Blank filling

根据对话内容，把下列句子补充完整，每空填写一个单词。

1. ________ is the process of removing impurities and solid particles from the sugarcane juice.
2. The juice undergoes ________ to remove most of the water content, leaving behind a thick syrup.
3. The machine used to separate the sugar crystals from the remaining liquid is called a ________.

Task 4: Thinking training

根据对话内容，完成从甘蔗中提取糖的工艺流程图，并试着复述流程的具体步骤。

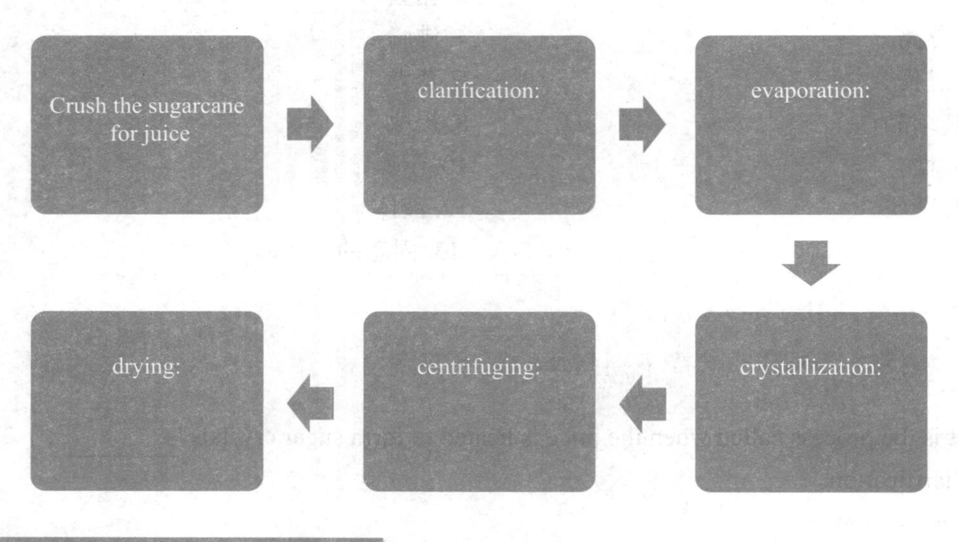

Key（Unit 2 Section 1）

Task 1:

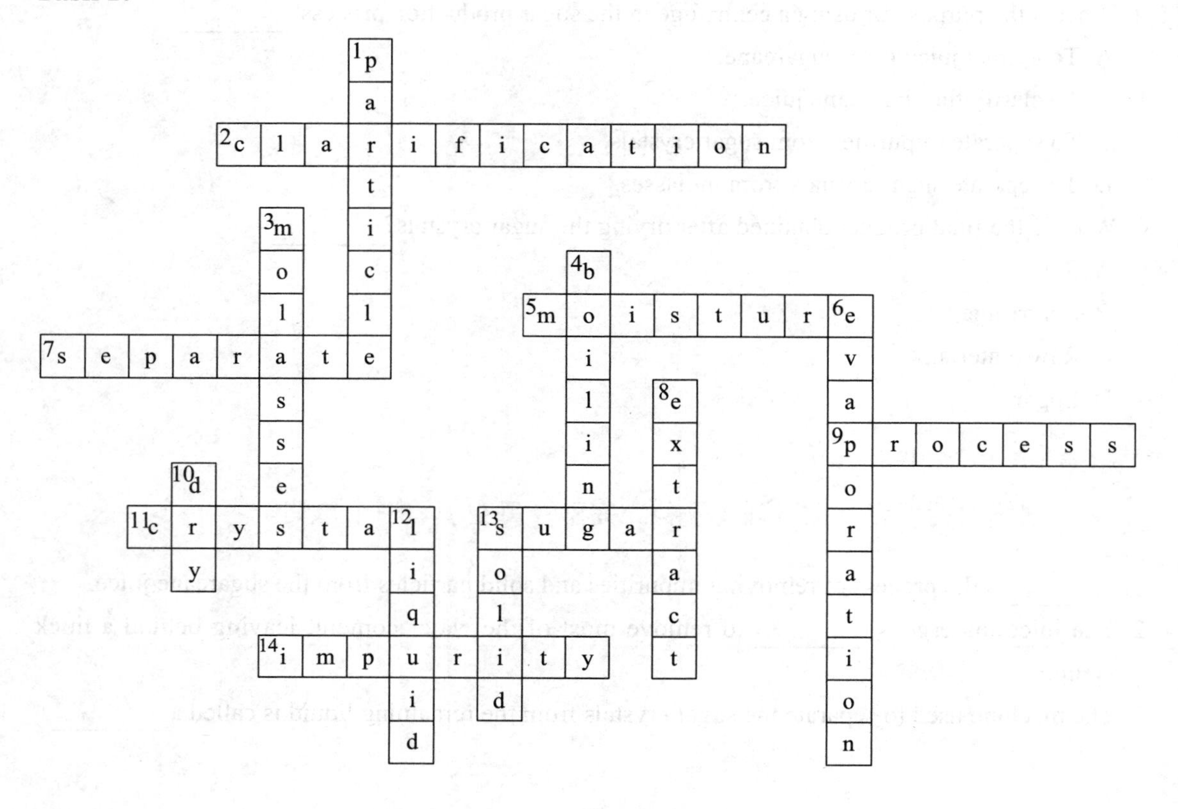

Task 2:

1. D 2. D 3. D

Task 3:

1. Clarification 2. evaporation 3. centrifuge

Task 4:

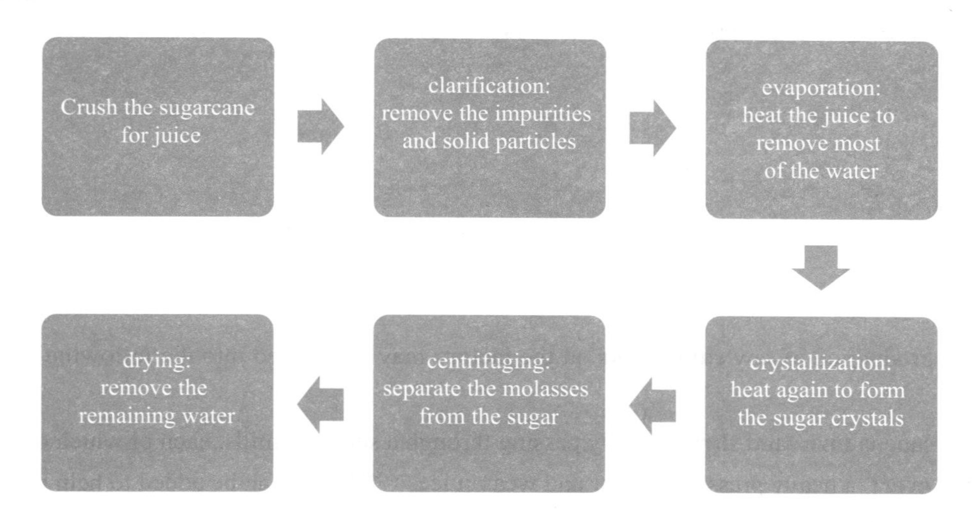

译文

第一节：从甘蔗中提取糖的过程

甜先生：同学们，早上好。今天，我们将讨论制糖的工作原理。让我们先来了解一下从甘蔗中提取糖的过程。

唐千禧：提取过程是如何进行的？

甜先生：首先要把甘蔗压碎，提取甘蔗汁。这个提取步骤对于获得制糖原料至关重要。

林华华：接下来会发生什么？

甜先生：榨汁之后，要经过一个澄清的过程。包括去除果汁中的杂质和固体颗粒，确保蔗汁清澈干净的状态。

张结晶：澄清后会发生什么？

甜先生：下一步是蒸发。在这个过程中，蔗汁被加热以除去大部分水分，留下一种黏稠的糖浆，我们称为糖蜜。

王提纯：那我们怎么从糖蜜中提取糖呢？

甜先生：把糖蜜煮沸，再次加热形成糖晶体。这个过程被称为糖的结晶化或煮沸。

唐千禧：现在我们有糖晶体了。接下来怎么做？

甜先生：我们需要把糖晶体从剩余的液体中分离出来。这是通过离心机完成的，离心机是一种高速旋转的机器，可以将糖晶体从糖浆中分离出来。这一步被称为离心法或从糖蜜分离糖。

林华华：太神奇了！我们终于得到纯糖晶体了？

甜先生：差不多了。最后一步是干燥糖晶体，去除剩余的水分。干燥完成，我们就得到了最终产品——糖！

张结晶：哇！这复杂的制糖过程真是令人惊叹。

甜先生：的确，从甘蔗变成我们都喜欢的甜晶体，这是一个奇妙的旅程。请记住，理解这个生产过程让我们感激生产日常用品所付出的努力，尽管我们通常认为这是理所当然的。

Section 2　Steps in the making of raw cane sugar

The production of raw **cane sugar**[1] at the factory may be divided into the following operation units.

The cane is **torn**[2] and **shredded**[3] by passing through a series of mills, each of which consists of three that **exert**[4] a heavy pressure. Water and weak juices（稀糖汁）can be added to help **macerate**[5] the cane and aid in（协助）the extraction. About 95 percent of the juice is extracted from the cane. The spent cane (**bagasse**[6]) is either burned for fuel or used to manufacture insulating material（绝缘材料）.

The juice is **screened**[7] to remove floating impurities and treated with **lime**[8] to **coagulate**[9] part of the **colloidal**[10] matter, **precipitate**[11] some of the impurities, and change the **pH**[12]. Phosphoric acid（磷酸） can be added because juices with a small amount of phosphates（磷酸盐）clarifies well. If acid is added, an excess of lime is used. The mixture is heated with high-pressure steam and settled in large tanks called clarifiers or **thickeners**[13].

To recover the sugar from the settled-out muds continuous rotary drum vacuum filters （连续转鼓真空吸滤机）are generally used although plate-and-frame presses（板框压滤机）may be employed. The cake constitutes 1 to 4 percent of the weight of cane charged and can be used as manure.

The filtrate（过滤液）, a clarified juice of high lime content, contains about 85 percent water. It is evaporated to approximately（大约）40 percent water in triple or quadruple-effect （三效或四效）**evaporators**[14]. The resulting dark-brown, viscous liquid （黏稠液体）is **crystallized**[15] in single-effect vacuum pan（单效真空结晶罐）. The syrup is concentrated until crystal formation begins.

The mixture of syrup and crystals (**massecuite**[16]) is **dumped into**[19] a **crystallizer**[17] which is a horizontal agitated tank （卧式搅拌槽）**equipped with**[20] cooling coils（冷却线圈）. Here additional sucrose deposits on（沉积）the crystals already formed, and crystallization is completed.

The massecuite is **centrifugated**[18] to remove the syrup. The crystals are high-grade **raw sugar**[21], and the syrup is retreated to obtain more crystals. The final liquid after reworking is known as **blackstrap molasses**[22].

The raw sugar (light brown in color), containing approximately 97 percent sucrose, is packed in bags and shipped to the refinery（糖厂）.

The molasses is used as a source of carbohydrates（碳水化合物）for fermentation（发酵）and for cattle feed.

音频：Unit 2　Section 2

New words and expressions

1. cane sugar	/keɪn ʃʊgər/	*n.* 蔗糖
2. torn	/tɔːn/	*v.* 撕碎，撕裂
3. shred	/ʃred/	*v.* 切碎，撕碎
4. exert	/ɪg'zɜːt/	*v.* 运用，施加（压力）
5. macerate	/'mæsəreɪt/	*v.* 浸软
6. bagasse	/bə'gæs/	*n.* 甘蔗渣
7. screened	/skriːnd/	*adj.* 筛过的
8. lime	/laɪm/	*n.* 石灰
9. coagulate	/kəʊ'ægjuleɪt/	*v.* 凝结
10. colloidal	/ˌkɒ'lɔɪdəl/	胶体的
11. precipitate	/prɪ'sɪpɪteɪt/	*v.* 沉淀
12. pH		酸碱度
13. thickener	/'θɪkənə(r)/	*n.* 稠化器
14. evaporator	/ɪ'væpəreɪtə(r)/	*n.* 蒸发器
15. crystallize	/'krɪstəlaɪz/	*v.*（使）结晶
16. massecuite	/mæ'skwiːt/	*n.* 糖膏
17. crystallizer	/'krɪstəlaɪzə(r)/	*n.* 结晶器；结晶设备
18. centrifugate	/'sentrɪfjugeɪt/	*v.* （使）离心
19. dump into		倾倒进
20. equip with		配备
21. raw sugar		原蔗糖
22. blackstrap molasses		黑糖蜜；粗炼糖蜜

Exercises

Task 1: Crosswords puzzle

请根据中文提示把下面字谜中的单词填写出来。

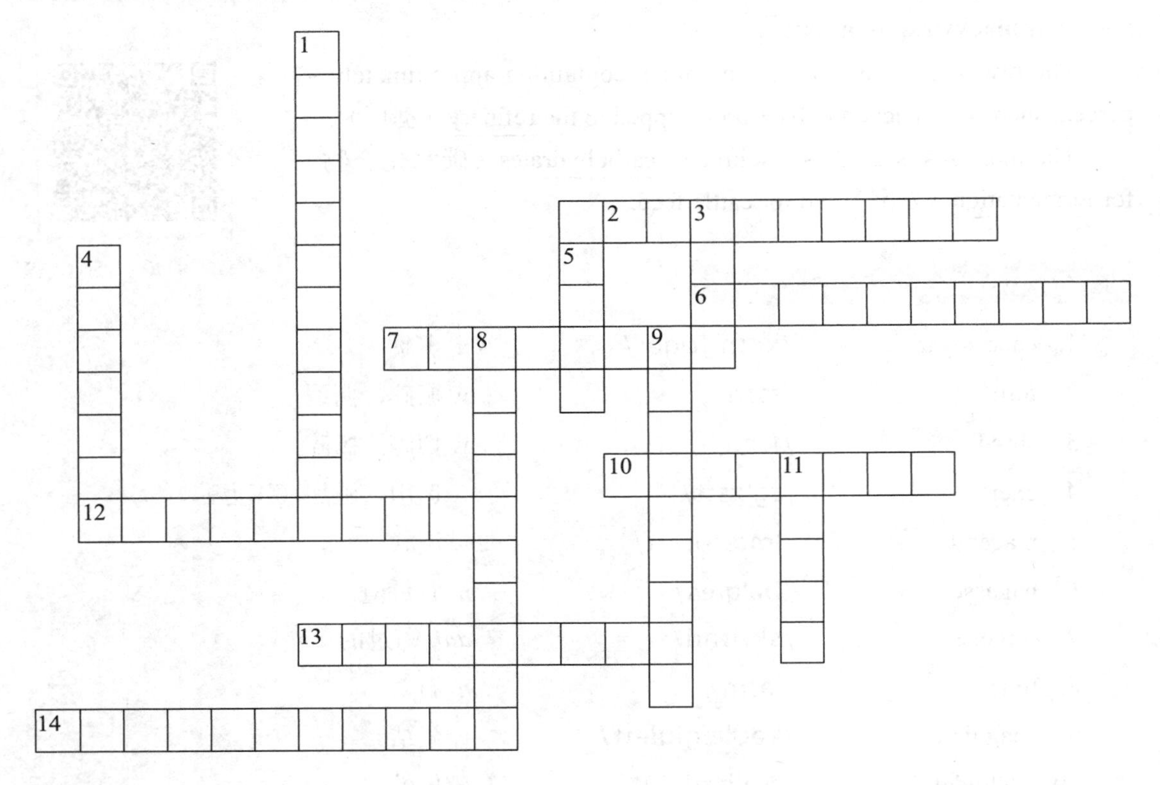

Across（横排）	**Down（竖排）**
2. 胶体的	1. 结晶器
6. 糖膏	3. 石灰
7. 浸软	4. 甘蔗渣
10. 筛过的	5. 撕碎
12. 蒸发器	8. 离心机
13. 凝结	9. 稠化器
14. 沉淀	11. 施加（压力）

Task 2: One-choice question

根据文章内容，选择下列问题的正确答案。

1. What is the purpose of passing the crushed cane through a series of mills?________

   A. To remove impurities from the cane.　　B. To extract juice from the cane.

   C. To burn the cane for fuel.　　D. To manufacture insulating material.

2. What is added to the juice to help coagulate impurities?________

A. Lime.　B. Water.　C. Weak juices.　D. Phosphoric acid.

3. What is the final product after the syrup and crystals are dumped into a crystallizer?________

A. Raw sugar.　B. Bagasse.

C. Blackstrap molasses.　D. Juice.

4. What is the main use of molasses?________

A. Source of carbohydrates for fermentation.　B. Cattle feed.

C. Manufacturing insulating material.　D. To extract sucrose.

Task 3: Blank filling

根据文章内容，把下列句子补充完整，每空填写一个单词。

1. The juice is ________ to remove floating impurities.
2. The mixture of syrup and crystals is dumped into a ________.
3. After the massecuite is centrifuged, the crystals are high-grade ________.
4. The raw sugar is packed in bags and shipped to the ________.

Task 4: Thinking training

根据文章内容，把下列关于制作原蔗糖的表格填写完整，并尝试与搭档一起简要描述制作原蔗糖的步骤。

Steps步骤	Results效果
Step 1: · Passing the crushed cane through a series of mills.	To________juice from the cane.
Step 2: · Screening the juice. · Adding________to the juice. · Heating the mixture with high-pressure steam and settled in thickener.	To remove floating________. To help coagulate impurities. To________some of the impurities.
Step 3: · Using continuous rotary drum vacuum filters （连续转鼓真空吸滤机）.	To recover the sugar from the settled-out muds.
Step 4: · Evaporating the filtrate（过滤液）.	To get the________, viscous liquid （黏稠液体）. The liquid is________in single-effect vacuum pan（单效真空结晶罐）
Step 5: · The mixture of syrup and crystals (massecuite) is dumped into a________.	The crystallization is completed.

Steps步骤	Results效果
Step 6: · Centrifuging the massecuite.	To remove the syrup. Then get the ________, that is high-grade raw sugar. The final liquid after reworking is known as ________.
Step 7: · The raw sugar is packed in bags and shipped to the refinery（糖厂）.	The color of the raw sugar is ________, containing about ________ percent sucrose.

Key（Unit 2　Section 2）

Task 1:

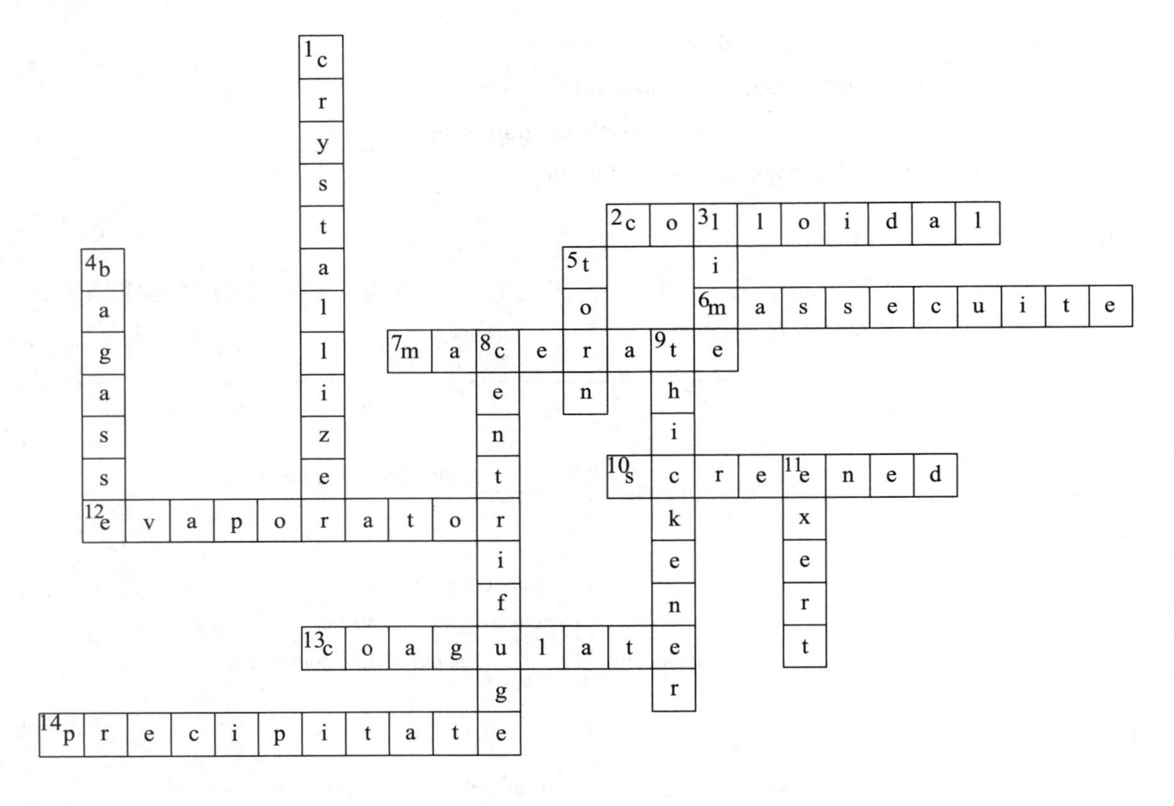

Task 2:

1. B　2. A　3. A　4. A

Task 3:

1.screened　2. crystallizer　3. raw sugar　4. refinery

Task4:

Steps步骤	Results效果
Step 1: · Passing the crushed cane through a series of mills.	To extract juice from the cane.
Step 2: · Screening the juice. · Adding lime to the juice. · Heating the mixture with high-pressure steam and settled in thickener.	To remove floating impurities. To help coagulate impurities. To precipitate some of the impurities.
Step 3: · Using continuous rotary drum vacuum filters（连续转鼓真空吸滤机）.	To recover the sugar from the settled-out muds.
Step 4: · Evaporating the filtrate（过滤液）.	To get the dark-brown, viscous liquid （黏稠液体）. The liquid is crystallized in single-effect vacuum pan（单效真空结晶罐）
Step 5: · The mixture of syrup and crystals (massecuite) is dumped into a crystallizer.	The crystallization is completed.
Step 6: · Centrifuging the massecuite.	To remove the syrup. Then get the crystals, that is high-grade raw sugar. The final liquid after reworking is known as blackstrap molasses.
Step 7: · The raw sugar is packed in bags and shipped to the refinery（糖厂）.	The color of the raw sugar is light brown, containing about 97 percent sucrose.

译文

第二节：制作原蔗糖的步骤

在工厂中，原蔗糖的生产可分为以下几个操作单元。

将甘蔗茎通过一系列的压榨机进行撕裂和切碎，每台压榨机都包括三个凹槽轧辊，施加着巨大的压力。可以添加水和稀糖汁使甘蔗浸软并协助萃取。大约有95%的蔗汁从甘蔗中提取出来。废弃甘蔗（蔗渣）要么被燃烧用作燃料，要么被用来制造绝缘材料。

蔗汁经筛选去除浮游杂质，并用石灰处理以凝固部分胶体物质、沉淀一些杂质并改变pH值（酸碱度）。可以添加磷酸，因为含少量磷酸盐的蔗汁很好澄清。如果添加了酸，就会使用过量的石灰。将混合物用高压蒸汽加热，然后在称为澄清器、连续沉降器或稠化器的大型储槽中沉淀。

为了从沉淀浆中回收糖，通常使用连续转鼓真空吸滤机，也可以使用板框压滤机。筛饼约占所加甘蔗重量的1%至4%，可用作肥料。

过滤液是高石灰含量的澄清蔗汁，含水量约为85%。在三效或四效蒸发器中将其蒸发至约40%的含水量。所得的深褐色黏稠液体在单效真空结晶罐中结晶。糖浆浓缩到结晶开始形成的程度。

糖浆和结晶物（糖膏）被倾倒入结晶器的卧式搅拌槽，该结晶器配备冷却线圈。这里，额外的蔗糖沉积在已经形成的结晶上，结晶过程完成。

然后，将糖膏离心以去除糖浆。晶体是高品质的原蔗糖，而糖浆则经再处理以获得更多的糖结晶。再处理后的最终液体被称为黑糖蜜。

原蔗糖（颜色为浅棕色）含有约97%的蔗糖，被装入袋子中并运往糖厂。

糖蜜被用作发酵的碳水化合物来源和牲畜饲料。

Culture Background

The Source of Sweetener in Ancient China

The first sweetener produced in northern China was maltose made from grains. Around the 2nd century BC, possibly even earlier, in many parts of northern China, mass production of maltose began. Nowadays, maltose made from barley and sorghum is still being produced in many regions of northern China. But in the 3rd century BC, sugarcane was most likely introduced from South Asia to southern China. Since then, sugarcane, as the main source of sweetener, has played a very important role in the history of Chinese food development.

古代中国甜料的来源

中国北方最初生产的甜料是由谷物制成的麦芽糖。大约在公元前2世纪，可能更早，在中国北方的很多地方，已开始大量生产麦芽糖。如今，在中国北方很多地区仍在生产由大麦和高粱制成的麦芽糖。但公元前3世纪，甘蔗很有可能是从南亚，引入中国南部，从那以后，甘蔗作为甜料的主要来源，在中国食品发展史上扮演着十分重要的角色。

Unit 3 Extraction in Sugarcane Mill

Objectives

After learning this unit, you will be able to

1. Expand your vocabulary about sugar cane mill and know the working process;
2. Summarize the working process of the sugar cane mill.

Section 1 Thinking about how many parts of a sugarcane mill and how do they work

Characters:
Mr. Sweet—a sugar industry professional （制糖行业的专业人士）
Lin Huahua—a student majoring in Sugar Engineering and being interested in sugar production (糖业工程专业学生，对糖的生产过程有浓厚兴趣)

Mr. Sweet: Good morning, class! Today, we are going to talk about the **sugarcane mill**[1] used in a **sugar factory**[2]. Lin Huahua, can you tell us about the different parts of the mill?

Lin Huahua: Sure! The sugarcane mill, also known as a mill, has several important **components**[3]. The **top roller**[4], or the "Ding gun" in Chinese, is **responsible for**[5] **pressing**[6] the sugarcane and extracting the **juice**[7]. It is **positioned**[8] above the other rollers.

Mr. Sweet: Excellent! And what about the other rollers?

Lin Huahua: The **feed roller**[9], or the "Qian gun" in Chinese, is the roller that feeds the sugarcane into the mill. It is located just below the top roller. The **discharge roller**[10], or the "Hougun" in Chinese, is the roller that receives the **crushed**[11] cane after the juice has been extracted. It is positioned below the feed roller.

Mr. Sweet: Very well **explained**[12]! Is there any other **vital**[13] part of the mill?

Lin Huahua: Yes, there is one more **crucial**[14] component called the **trash plate**[15], or the "di shu in Chinese". It is located at the bottom of the mill and is responsible for separating the cane fibers from the juice.

Mr. Sweet: Great job, Lin Huahua! You have explained the different parts of a sugar cane mill very clearly. Keep up the good work, everyone!

音频：Unit 3　Section 1

New words and expressions

1. sugarcane mill		压蔗机
2. sugar factory		糖厂
3. component	/kəmˈpəʊnənt/	*n.* 组件，构件
4. top roller		顶辊
5. be responsible for		负责；担负
6. press	/pres/	*v.* 压榨，挤
7. juice	/dʒuːs/	*n.* 汁液；果汁
8. position	/pəˈzɪʃ(ə)n/	*v.* 安置，处于
9. feed roller		前辊
10. discharge roller		后辊
11. crushed	/krʌʃt/	*adj.* 压碎的
12. explain	/ɪkˈspleɪn/	*v.* 解释；说明
13. vital	/ˈvaɪt(ə)l/	*adj.* 必不可少的；对……极重要的
14. crucial	/ˈkruːʃ(ə)l/	*adj.* 至关重要的；关键性的
15. trash plate		底梳

Exercises

Task 1: Crosswords puzzle

请根据中文提示把下面字谜中的单词填写出来。

Across（横排）

3. 压碎的
6. 辊，滚轴
8. 负责；担负
9. 至关重要的；关键的
12. 分离；区分；隔开区别

Down（竖排）

1. 安置，处于
2. 压榨，挤
4. 解释；说明
5. 必不可少的；对……极重要的
7. 提取
10. 组件，构件
11. 汁液

Task 2: One-choice question

根据对话内容，选择下列问题的正确答案。

1. What is the purpose of the top roller in a sugar cane mill?________

   A. It feeds the sugar cane into the mill.

   B. It separates cane fibers from the juice.

   C. It presses the cane and extracts the juice.

2. Where is the trash plate located in a sugar cane mill?________

A. Above the top roller.

B. Between the feed roller and the discharge roller.

C. At the bottom of the mill.

3. What does the feed roller do in a sugar cane mill?________

A. It extracts the juice.

B. It feeds the sugar cane into the mill.

C. It separates the cane fibers.

Task 3: Blank filling.

根据对话内容，把下列句子补充完整，每空填写一个单词。

1. The ________ roller receives the crushed cane after the juice has been extracted.
2. The sugar cane mill is responsible for ________ the cane fibers from the extracted juice.
3. The trash plate is located at the ________ of the mill.

Task 4:Thinking training

根据对话内容，把压蔗机部件、所处位置用流程图的形式绘制出来，并试着复述。

The working process of a sugarcane mill

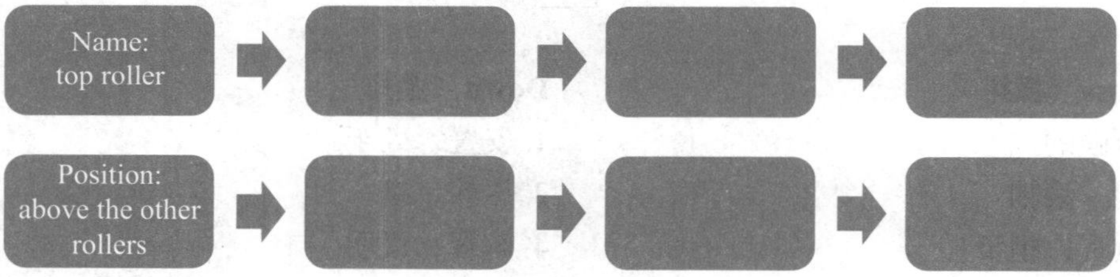

Name: top roller	→		→		→	
Position: above the other rollers	→		→		→	

____

____

____

____

____

____

____

____

____

____

Key (Unit 3 Section 1)

Task 1:

Task 2:

1. C 2. C 3. B

Task 3:

1. discharge 2. separating 3. bottom

Task 4:

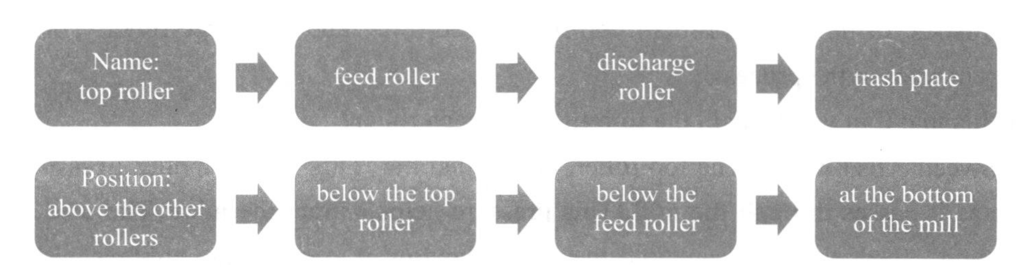

The sugarcane mill has several important parts. The top roller (Ding gun) is positioned above the other rollers and is responsible for pressing the sugarcane and extracting juice. The feed roller (Qian gun) is located just below the top roller and feeds the sugarcane into the mill. The discharge roller (Hougun) is positioned below the feed roller and receives the crushed cane after juice extraction. The trash plate (di shu) is located at the bottom of the mill and is responsible for separating cane fibers from the juice.

译文

第一节：压蔗机的组成与运作

甜先生：早上好，同学们！今天，我们来说说糖厂中使用的压蔗机。林华华，你能给我们介绍一下压蔗机的组成吗？

林华华：可以！压蔗机，也称为蔗机，有几个重要的组成构件。首先是“顶辊”，负责压榨甘蔗和提取蔗汁，位于所有辊的上方。

甜先生：讲得太好了！那么其他辊又有哪些呢？

林华华：其次是“前辊”，是将甘蔗送榨的辊。它位于顶辊的正下方。接下来是后辊，在压榨后接收碎甘蔗。它位于前辊下方。

甜先生：解释得很好！还有其他重要部件吗？

林华华：是的，还有一个重要部件，叫做“底梳”。它位于压蔗机的底部，负责从蔗汁中分离甘蔗渣。

甜先生：说得好，林华华！你已经非常清楚地解释了压蔗机的组成。其他同学也要努力！

Section 2 The working principle of a sugarcane mill

Mr. Sweet: Good morning, class! Today, we are going to learn about the working **principle**[1] of a sugarcane mill. Let's dive right in!

Tang Qianxi: What does a mill do **exactly**[2]?

Mr. Sweet: Great question! A sugarcane mill is **designed**[3] to extract juice from the sugarcane. It **consists**[4] of three main rollers: top roller, feed roller and discharge roller. These rollers play different roles in the process.

Lin Huahua: Can you explain the **functions**[5] of each roller?

Mr. Sweet: **Absolutely**[6]! The top roller, also called the "Ding gun", applies **pressure**[7] to the sugarcane as it passes through the mill. This pressure helps extract the juice. The feed roller, or the "Qian gun", feeds the sugarcane into the mill, **allowing**[8] it to come into **contact with**[9] the top roller. Finally, the discharge roller, or the "Hou gun", receives the crushed cane after the juice has been extracted.

Zhang Jiejing: What happens to the sugarcane during this process?

Mr. Sweet: Excellent question! To separate the cane fibers from the juice by trash plate, or "Di shu" is used. The trash plate is positioned at the bottom of the mill. As the sugarcane passes through the mill, it **encounters**[10] the trash plate, which allows the juice to **flow**[11] through while **retaining**[12] the fibers. This **separation**[13] process ensures that we **obtain**[14] clean cane juice.

Wang Tichun: So, the end result is the juice, right?

Mr. Sweet: Yes, exactly! The mill's **primary**[15] **objective**[16] is to extract juice, also known as "Zhe zhi" from the sugarcane. This juice is further processed to produce sugar and other sugary products. Remember, the mill's rollers and the trash plate work together to extract the juice and separate the fibers, allowing us to obtain pure sugarcane juice.

音频：Unit 3　Section 2

New words and expressions

1. principle	/'prɪnsəp(ə)l/	*n.* 工作原理；法则
2. exactly	/ɪg'zæk(t)li/	*adv.* 正是；完全正确；究竟
3. design	/dɪ'zaɪn/	*v.* 设计；构思
4. consist	/kən'sɪst/	*v.* 构成；包含
5. function	/'fʌŋkʃ(ə)n/	*n.* 功能；作用
6. absolutely	/'æbsəlu:tli/	*int.* 当然
7. pressure	/'preʃə(r)/	*n.* 压力；压强；挤压
8. allow	/ə'laʊ/	*v.* 让；允许
9. contact with		与……接触；与……联系
10. encounter	/ɪn'kaʊntə(r)/	*v.* 遇到；相遇
11. flow	/fləʊ/	*v.* 流；流过
12. retain	/rɪ'teɪn/	*v.* 保留；保存；维持
13. separation	/sepə'reɪʃ(ə)n/	*n.* 分离；分开
14. obtain	/əb'teɪn/	*v.* 获得；得到
15. primary	/'praɪməri/	*adj.* 主要的；最重要的
16. objective	/əb'dʒektɪv/	*n.* 目标；目的

Exercises

Task 1: Crosswords puzzle

请根据中文提示把下面字谜中的单词填写出来。

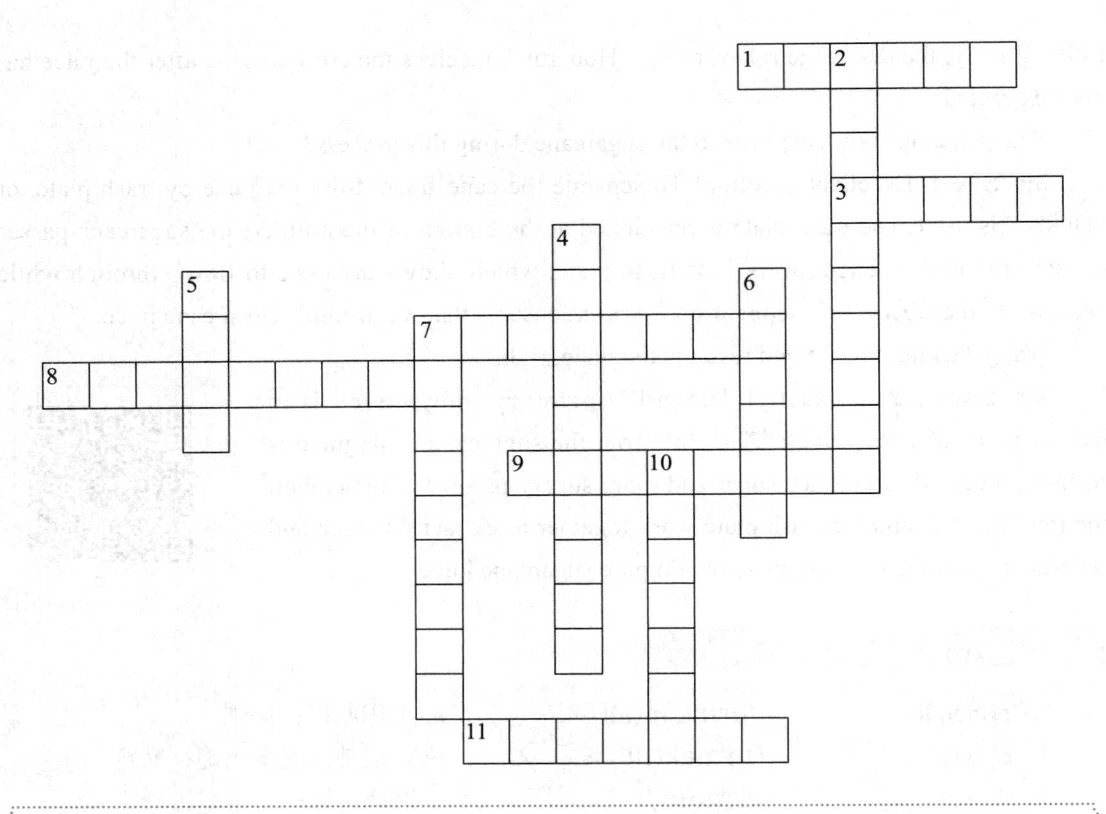

Across（横排）

1. 设计；构思
3. 让；允许
7. 压力；压强；挤压
8. 遇到；相遇
9. 功能；作用
11. 正是；完全正确；完美

Down（竖排）

2. 分离；分开
4. 当然
5. 流；流过
6. 保留；保存；维持
7. 工作原理；法则
10. 构成；包含

Task 2: One-choice question

根据对话内容，选择下列问题的正确答案。

1. What is the primary purpose of a sugarcane mill?________

   A. To produce sugarcane.　　B. To extract cane juice.

   C. To remove cane fibers.

2. The top roller in a sugarcane mill applies ________ to extract juice.

   A. heat　　B. pressure　　C. grinding

3. What is the role of the trash plate in a sugarcane mill?________

   A. To separate cane fibers from juice.　　B. To feed cane into the mill.

   C. To discharge crushed cane.

Task 3: Blank filling

根据对话内容，把下列句子补充完整，每空填写一个单词。

1. Sugarcane mills are designed to ________ juice from the sugarcane.
2. The ________ roller feeds the sugarcane into the mill.
3. The ________ roller receives the crushed cane after the juice has been extracted.

Task 4: Reading comprehension

根据对话内容，回答下列问题。

1. According to the dialogue, what are the functions of each roller?

____

____

2. How does the trash plate work?

____

____

Key (Unit 3 Section 2)

Task 1:

Task 2:

1. B 2. B 3. A

Task 3:

1. extract 2. feed 3. discharge

Task 4:

1. The top roller applies pressure to the sugarcane as it passes through the mill. This pressure helps extract the juice. The feed roller feeds the sugarcane into the mill, allowing it to come into contact with the top roller. The discharge roller receives the crushed cane after the juice has been extracted.

2. The trash plate is used to separate the cane fibers from the juice. The trash plate is positioned at the bottom of the mill. As the cane passes through the mill, it encounters the trash plate, which allows the juice to flow through while retaining the fibers.

译文

第二节：压蔗机的工作原理

甜先生：早上好，同学们！今天，我们将了解压蔗机（也称为蔗机）的工作原理。话不多说，我们马上开始！

唐千禧：压蔗机究竟是做什么用的？

甜先生：问得好！压蔗机的设计用于从甘蔗中提取甘蔗汁。它主要由三个辊组成：顶辊、前辊和后辊。这些辊在这个过程中起着不同的作用。

林华华：能解释一下每个辊的功能吗？

甜先生：当然！“顶辊”在甘蔗通过蔗机时对甘蔗施加压力以提取甘蔗汁。前辊将甘蔗送入蔗机，让其与顶辊接触。最后，后辊在提取蔗汁后接收压碎的甘蔗。

张结晶：在这个过程中，甘蔗会怎样变化？

甜先生：很好的问题！我们使用底梳将甘蔗纤维与甘蔗汁分离。底梳于蔗机底部，当甘蔗通过蔗机时，它会遇到底梳，它让蔗汁流过，同时保留住了纤维。这种分离过程确保我们获得纯净的甘蔗汁。

王提纯：所以，最终的结果就是获取甘蔗汁吧？

甜先生：正是！甘蔗机的主要目标就是从甘蔗中提取甘蔗汁，也称为“蔗汁”。得到的蔗汁被进一步加工以生产糖和其他含糖产品。请记住，蔗机的辊和底梳一起工作以提取蔗汁并分离纤维，使我们能够获得纯净的甘蔗汁。

Section 3　Talking about the working process of a mill workshop

Lin Huahua: Hey, have you read the **article**[1] about the **mill workshop**[2]?

Tang Qianxi: Yeah, I found it quite interesting. Did you know that the sugarcanes are first cut into short pieces by the **cutter**[3]?

Lin Huahua: Yes, and then they pass through the **three-roll crushers**[4] to extract the juice. It's **fascinating**[5] how the rolls have **corrugations**[6].

Tang Qianxi: Absolutely! And **depending on**[7] whether water is **added**[8] or not during crushing, the canes are either **dry-crushed**[9] or **wet-crushed**[10].

Lin Huahua: Right, and the juice is collected in **troughs**[11], warmed to a **specific**[12] **temperature**[13], and then **settled**[14] in **tanks**[15] for a short time. It's quite a **meticulous**[16] process.

Tang Qianxi: Definitely. I also found it **intriguing**[17] that the spent bagasse can be used as fuel or as a source of **cellulose**[18] for various purposes.

Lin Huahua: It's amazing how the amount of juice extracted from the canes varies with different **factors**[19], like the maturity of the canes and the equipment used.

Tang Qianxi: Yes, it just goes to show how **complex**[20] and **precise**[21] the sugarcane extraction process is.

Lin Huahua: Absolutely. I'm glad we had a chance to discuss this article. It's always interesting to learn about different industries and processes.

Tang Qianxi: I couldn't agree more. Learning new things is always a **rewarding**[22] experience.

音频：Unit 3　Section 3

New words and expressions

1. article	/ˈɑː(r)tɪk(ə)l/	*n.* 文章
2. mill workshop		压榨车间
3. cutter	/ˈkʌtə(r)/	*n.* 刀具，切割机
4. three-roll crushers		三辊压榨机
5. fascinating	/ˈfæsɪneɪt/	*adj.* 吸引人的
6. corrugation	/ˌkɒrʊˈgeɪʃən/	*n.* 沟纹
7. depend on		取决于；视……而定
8. add	/æd/	*v.* 添加；增加
9. dry-crushed		干榨
10. wet-crushed		湿榨
11. trough	/trɒf/	*n.* 槽形容器

12. specific	/spə'sɪfɪk/	*adj.* 特定的；具体的
13. temperature	/'temprɪtʃə(r)/	*n.* 温度；气温
14. settle	/‹set(ə)l/	*v.* 沉淀；停留
15. tank	/tæŋk/	*n.* 罐
16. meticulous	/mɪ'tɪkjʊləs/	*adj.* 一丝不苟的；细致的
17. intriguing	/ɪn'triːgɪŋ/	*adj.* 非常有趣的；引人入胜的
18. cellulose	/'seljuləʊs/	*n.* 纤维素
19. factor	/'fæktə(r)/	*n.* 因素；要素
20. complex	/'kɒmpleks/	*adj.* 复杂的；难懂的
21. precise	/prɪ'saɪs/	*adj.* 准确的；确切的；精确的
22. rewarding	/rɪ'wɔː(r)dɪŋ/	*adj.* 值得做的；有益的

Exercises

Task 1: Crosswords puzzle

请根据中文提示把下面字谜中的单词填写出来。

Across（横排）

2. 纤维素

3. 槽形容器

Down（竖排）

1. 特定的；具体的

3. 温度；气温

4. 文章
5. 刀具，切割机
6. 甘蔗渣
7. 添加；增加
8. 罐
9. 吸引人的
10. 沉淀；停留

Task 2: One-choice question

根据对话内容，选择下列问题的正确答案。

1. The sugarcanes are first cut by ________ .

   A. the cutter
   B. three-roll crushers
   C. corrugations
   D. the mill

2. Which part is used to extract the cane juice?________

   A. The troughs.
   B. The three-roll crushers.
   C. The tanks.
   D. The cutter.

3. How can the spent bagasse be used?________

   A. They can be used as fuels.
   B. They can be used as a source of cellulose.
   C. They can be used for various purposes.
   D. All of the above.

Task 3: Blank filling

根据对话内容，把下列句子补充完整，每空填写一个单词。

1. The canes are first cut into short pieces by the ________ .
2. The spent bagasse can be used as ________ or as a source of cellulose.
3. The amount of ________ extracted from the canes varies with different factors.

Task 4: Thinking training

根据下列提示复述压榨车间的生产过程。

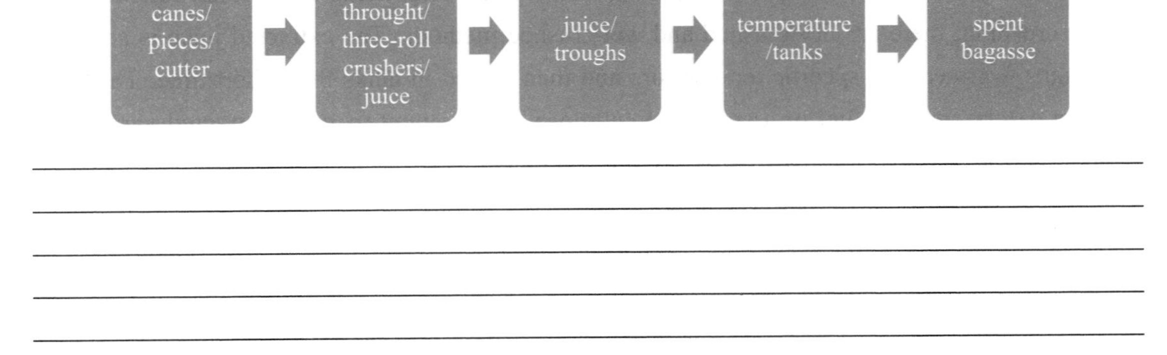

____

____

____

____

____

Key (Unit 3 Section3)

Task 1:

Task 2:

1. A 2. B 3. D

Task 3:

1. cutter 2. fuel 3. juice

Task 4:

In the mill workshop, sugarcanes are first cut into short pieces by the cutter. Then they pass through three-roll crushers with corrugations to extract juice. Depending on whether water is added during crushing, there are dry-crushed and wet-crushed methods. The extracted juice is collected in troughs, warmed to a specific temperature and then settled in tanks for a short time. The spent bagasse can be used as fuel or a source of cellulose. The amount of juice extracted varies with factors such as the maturity of the canes and the equipment used.

译文

第三节：讨论压榨车间的工作过程

林华华：嘿，你看过关于压榨车间的文章吗？

唐千禧：看了，我觉得挺有意思的。你知道甘蔗首先被切割机切成短块吗？

林华华：是的，接着甘蔗块会通过三辊压榨机以提取蔗汁。辊子上的沟纹太吸引人了。

唐千禧：是呀！取决于在压榨过程中是否添加水，还分为干榨和湿榨两种方式。

林华华：嗯，蔗汁流在槽形容器中，加热到特定的温度，然后在罐中进行短时间的沉淀。这是一个相当细致的过程。

唐千禧：没错，更有趣的是，用过的甘蔗渣可以用作燃料，或作为纤维素的来源，用途广泛。

林华华：令人惊讶的是，从甘蔗中提取的蔗汁量受不同因素的影响，比如甘蔗的成熟度和使用的设备。

唐千禧：是的，这正好说明蔗汁提取过程有多复杂精确。

林华华：是的。我很高兴我们有机会讨论这篇文章，了解不同的行业和生产流程总是很有趣。

唐千禧：完全同意，学习新事物总是有益的体验。

Section 4 Mill workshop

At the mill workshop the canes are first cut into short pieces by the rapidly **revolving**[1] knives of the cutter, and then pass to one or more three-roll crushers, where the juice is pressed out. The rolls **revolve**[2] slowly, and have corrugations. In the usual form, one roll **surmounts**[3] the lower two rolls, And is carried from one lower roll to the second by the trash plate. If no water is added, the cane is said to be dry-crushed; if water is played on the cane at the second or third set of rollers, it is wet-crushed. The pressed cane is the "bagasse", which is treated further. The juice drops into troughs under the rollers and is strained（过滤）, warmed to 200 ℉ (93 ℃) and run into settling tanks for a short time. Crushing rolls for sugarcane, with typical turn plate arrangement.

The spent bagasse is a sponge-like material; it serves as **fuel**[4], sometimes with the addition of **crude oil**[5], or fuel oil; as a source of cellulose; and for making **celotex board**[6].

The amount of juice in the cane varies in different **territories**[7]; it varies also with the degree of maturity. Of the juice in the cane 60 to 80 percent is extracted, depending on the perfection of the **equipment**[8]. Table 3-1 are **analysis**[9] of the juice.

音频：Unit 3 Section 4

Table 3-1　Analysis of the juice

Water	Sucrose	Reducing sugar	Undetermined solids	Total
83.6	14.1	0.6	1.7	100.0

New words and expressions

1. revolving　/rɪ'vɒlvɪŋ/　*adj.* 旋转的
2. revolve　/rɪ'vɒlv/　*v.* 旋转
3. surmount　/sə'maʊnt/　*v.* 耸立于……之上
4. fuel　/'fjuːəl/　*n.* 燃料
5. crude oil　/kruːd ɔɪl/　*n.* 原油
6. celotex board　隔声板
7. territory　/'terətri/　*n.* 地区
8. equipment　/ɪ'kwɪpmənt/　*n.* 设备
9. analysis　/ə'næləsɪs/　*n.* 分析

Exercises

Task 1: Crosswords puzzle

请根据中文提示把下面字谜中的单词填写出来。

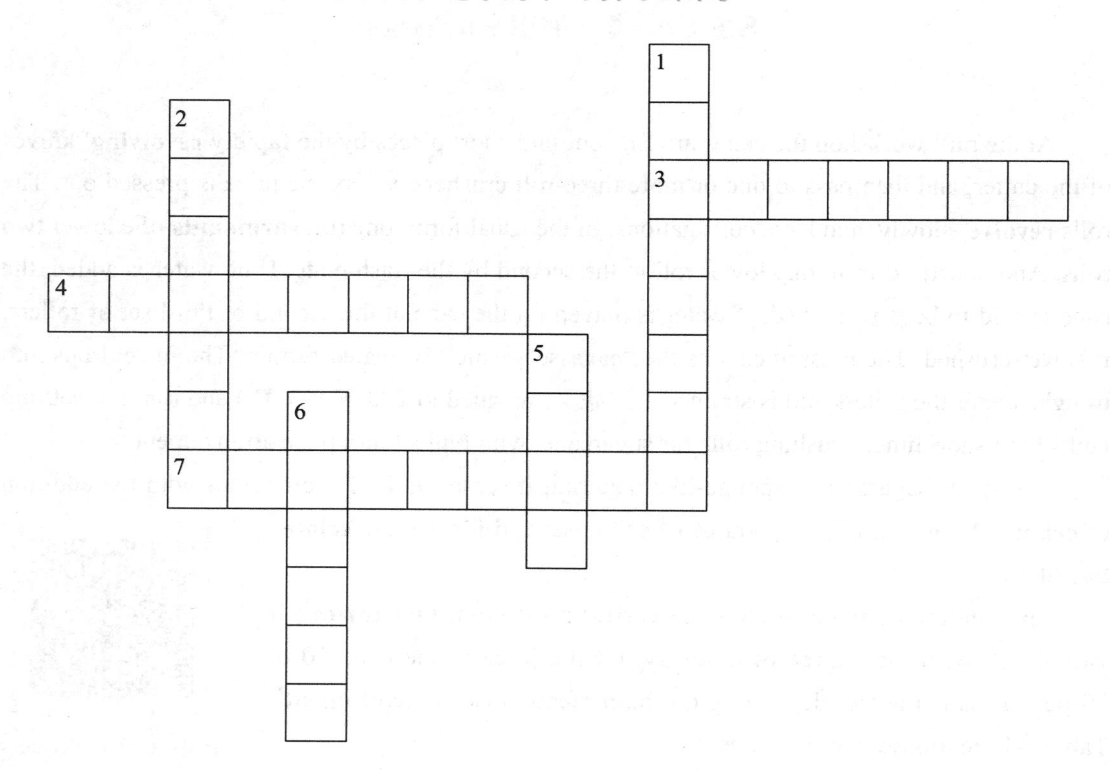

Across（横排）	**Down**（竖排）
3. 旋转	1. 耸立于……之上
4. 分析	2. 甘蔗渣
7. 设备	5. 燃料
	6. 切割机

Task 2: One-choice question

根据文章内容，选择下列问题的正确答案。

1. What is the purpose of the cutter in the mill house?________
   A. To press the juice out of the canes.
   B. To cut the canes into short pieces.
   C. To warm the juice to a specific temperature.
   D. To strain and settle the juice.
2. What is the difference between dry-crushed and wet-crushed canes?________
   A. The rollers used in crushing.
   B. The temperature of the juice.
   C. The amount of juice extracted.
   D. The speed of the revolving knives.
3. What is the main use of the spent bagasse?________
   A. Fuel.
   B. Source of cellulose.
   C. Making celotex board.
   D. Straining the juice.
4. What factors affect the amount of juice extracted from the canes?________
   A. The equipment used.
   B. The maturity of the canes.
   C. The temperature of the juice.
   D. The size of the cutter.

Task 3: Blank filling

根据文章内容，把下列句子补充完整，每空填写一个单词。

1. The juice is ________ out of the canes in the mill house.
2. The bagasse is used as fuel and a source of ________.
3. The amount of juice extracted from the canes depends on the ________ of the canes.
4. The juice is warmed to ________ and settled in tanks.

Task 4: Thinking training

根据文章内容，把下列关于压榨车间的设备名称及用途表格填写完整，并尝试与搭档一起描述甘蔗压榨的过程。

The equipment in the mill house	Usage
· ________ with the rapidly revolving knives	Cut the canes into short pieces
· Three-roll ________	The rolls ________ slowly, and pressed out the ________.
· Troughs · Settling tanks	The juice drops into ________ under the rollers and is strained（过滤）, warmed to 200 ℉(93 ℃) and run into ________ for a short time.

Key（Unit3 Section4）

Task 1:

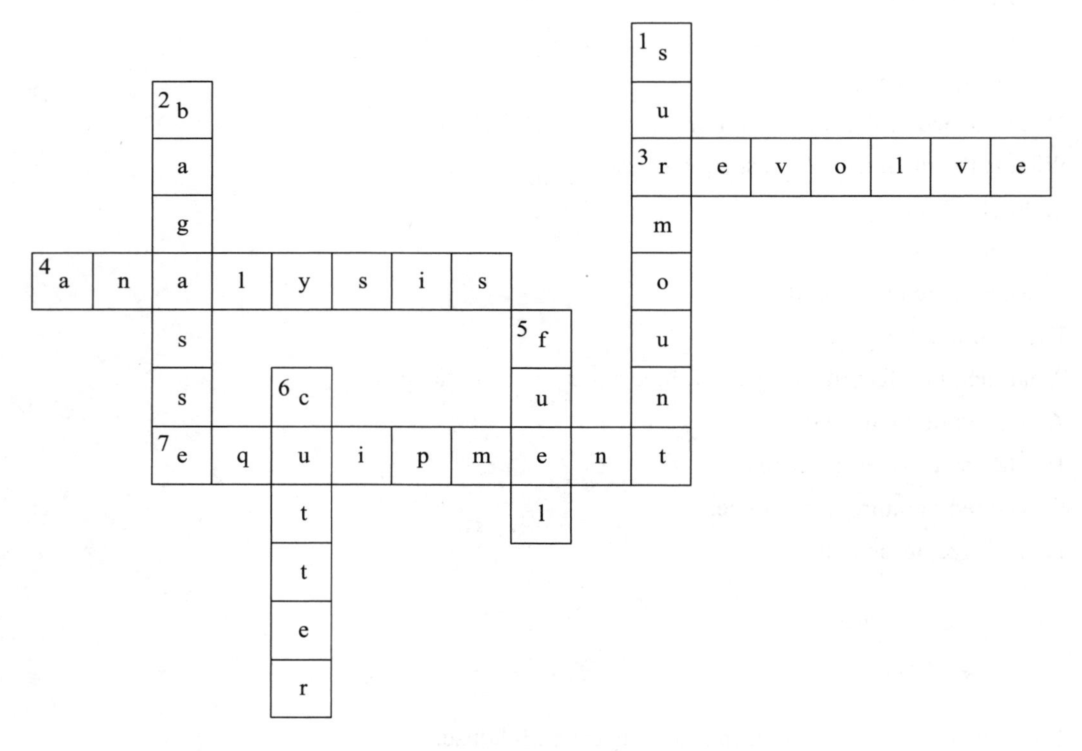

Task 2:

1. B 2. C 3. A 4. A

Task 3:

1. pressed 2. cellulose 3. maturity 4. 200 ℉ (93 ℃)

Task 4:

The equipment in the mill house	Usage
· Cutter with the rapidly revolving knives	Cut the canes into short pieces
· Three-roll crushers	The rolls revolve slowly, and pressed out the juice.
· Troughs · Settling tanks	The juice drops into troughs under the rollers and is strained（过滤）, warmed to 200 ℉(93 ℃) and run into settling tanks for a short time.

译文

第四节：压榨车间

在工厂里，甘蔗首先被快速旋转的蔗刀切成小段，然后进入一个或多个三辊压榨机，榨出汁液。压榨机缓慢旋转，压榨机轧辊上有沟纹。通常的样式是，一个辊架在下面的两个辊上，并由底梳从前辊运送到后辊。如果不加水，则称为干榨；如果在第二组或第三组轧辊上给蔗料上加水，则称为"湿榨"。被压榨后的甘蔗就是"甘蔗渣"，它会被进一步处理。蔗汁滴入辊下的蔗汁槽中，经过过滤，加热至 200 ℉（93 ℃），然后进入沉降池中快速沉降。甘蔗破碎辊采用典型的品字形布置。

用过的甘蔗渣是海绵状材料，可用作燃料，还可以添加到原油或燃料油中来使用；作为纤维素的来源；以及用于制造隔声板。

甘蔗中的蔗汁含量因地区而异，也因成熟度而异。根据设备的完善程度，甘蔗中 60% ~ 80% 的汁液被提取出来。表 3-1 对蔗汁的分析。

表 3-1 甘蔗汁成分表

水	蔗糖分	还原糖	未测定固溶物	总和
83.6	14.1	0.6	1.7	100.0

Section 5 Juice extraction

The cane is first cut into short lengths by rapidly revolving knives and then passed to a **milling train**[2] which consist, for example, of a two-roll crusher followed by four three-roll mills, 14 **rollers**[3] in all where the juice is **squeezed out**[4] by pressure（Fig.3-1）. The rolls in the three-roll mills are under a pressure of 75 to 80 tons per foot of roll length, and the length ranges from 2.5 to 7.0 feet; the **crushing**[5] **varies with**[6] the length of the roll, from 500 tons（吨）for the small length to 5000 tons per 24 hours for the larger mills（Fig.3-2）.

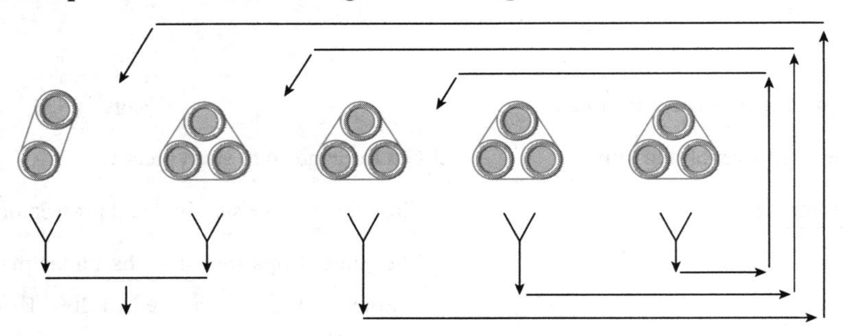

Fig.3-1 A milling train consisting of a crusher and four three-roll mills, grinding[7] with the application of water; compound[8] imbibition[9].

In the process of milling, water is applied to the crushed cane and the lean (**dilute**[10]) juices are returned to the mills. In a 14-roller milling train, water is applied to the cane at the third mill, and its juice is fed to the first mill; the juice from the fourth mill is fed to the second mill. The method is known as compound imbibition. The approximate（近似的）percentage of the sucrose in the cane extracted by each unit of the milling train is about as follows:

Table 3-2 The approximate percentage of the sucrose in the cane extracted by each unit of the milling train

Crusher	1st mill	2nd mill	3rd mill	4th mill	Total
70.0	12.5	7.0	4.5	2.0	96.0

Extraction[1] of cane juice by passing water counter-current to sliced cane is currently being investigated（研究）as a means of improving sugar **recovery**[11].

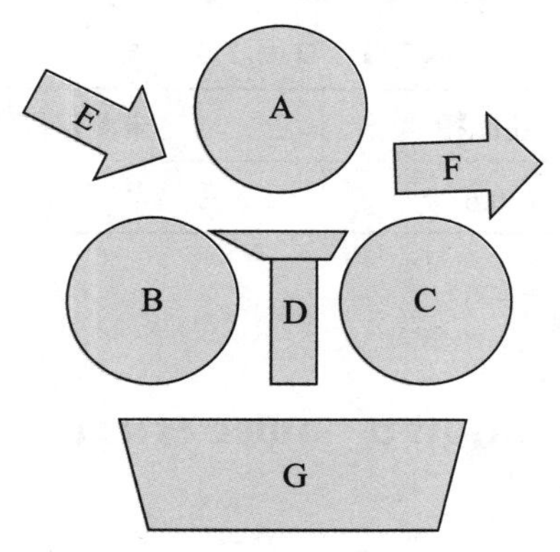

Fig.3-2 A cane mill with its three crushing rolls and with a typical turn-plate arrangement. An hydraulic piston（液压活塞）maintains pressure on A, while B and C are fixed; the rolls are under pressure of 75-80 tons per foot of roll width.

A—Top-roller; B—Bagusse roller; C—Feed roller; E—Feed of cane; F—Discharge of crushed cane; G—Juice-cdlecting though

The **composition**[12] of the juice varies with soil and climatic conditions, cane variety, and degree of maturity, within the limits indicated in Table 3-3:

Table 3-3 Ingredient list of sugar juice

Water	Sucrose	Sugars	Compounds	**Ash**[13]	pH
78%～86%	10%～20%	0.5%～2.5%	0.5%～1.0%	0.3%～0.7%	5.1～5.7

As shown by the pH, the juice is **acid**[14]; it is strained through a **perforated**[15] metal sheet and sent to the boiling workshop where it is clarified and **concentrated**[16] by boiling. The bagasse from the last mill of a train contains from 52 to 58 percent of solid matter, mainly cane fiber; the **remainder**[17] is water. It is used primarily as a fuel for raising the steam required by the factory.

音频：Unit 3 Section 5

New words and expressions

1. extraction	/ɪk'strækʃn/	*n.* 提取
2. milling train		压榨机组
3. roller	/'rəʊlə(r)/	*n.* 轧辊
4. squeeze out	/skwiːz aʊt/	挤出
5. crushing	/'krʌʃɪŋ/	轧碎
6. vary with		随……而变化
7. grind	/graɪnd/	*v.* 磨碎，碾碎
8. compound	/'kɒmpaʊnd/	*adj.* 复合的
9. imbibition	/ˌɪmbɪ'bɪʃən/	*n.* 渗浸
10. dilute	/daɪ'luːt/	*adj.* 稀释的
11. recovery	/rɪ'kʌvəri/	*n.* 回收
12. composition	/ˌkɒmpə'zɪʃ(ə)n/	*n.* 成分构成
13. ash	/æʃ/	*n.* 灰
14. acid	/'æsɪd/	*adj.* 酸性的
15. perforated	/'pɜːfəˌreɪtɪd/	*adj.* 穿孔的
16. concentrate	/'kɒns(ə)ntreɪt/	*v.* 浓缩
17. remainder	/rɪ'meɪndə(r)/	*n.* 剩余部分

Exercises

Task1: Crosswords puzzle

请根据中文提示把下面字谜中的单词填写出来。

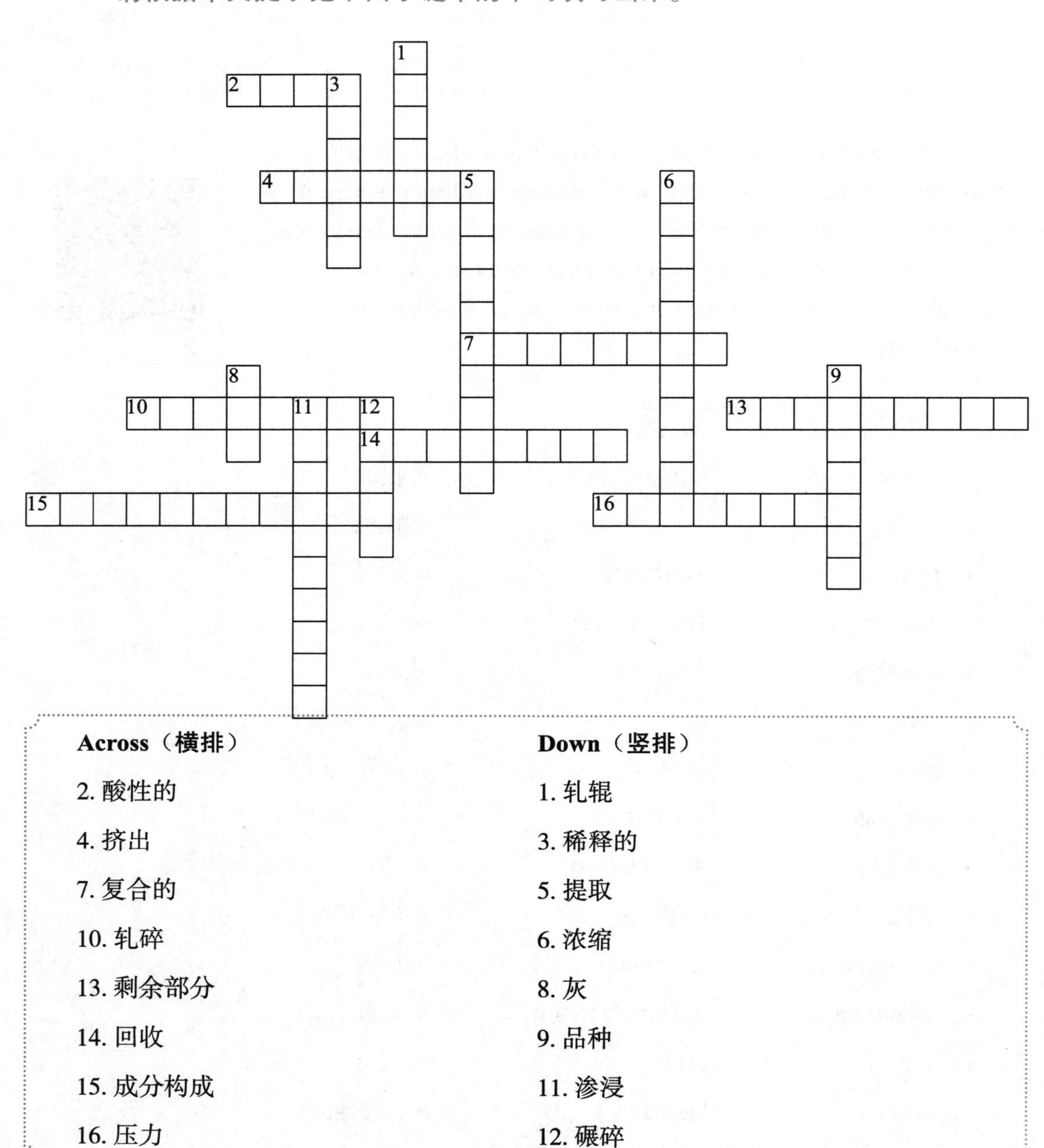

Across（横排）	Down（竖排）
2. 酸性的	1. 轧辊
4. 挤出	3. 稀释的
7. 复合的	5. 提取
10. 轧碎	6. 浓缩
13. 剩余部分	8. 灰
14. 回收	9. 品种
15. 成分构成	11. 渗浸
16. 压力	12. 碾碎

Task 2: One-choice question

根据对话内容，选择下列问题的正确答案。

1. What is the purpose of the milling train in juice extraction?________

   A. To cut the cane into short lengths.

   B. To squeeze out the juice through pressure.

   C. To apply water to the crushed cane.

   D. To improve sugar recovery.

2. What is the approximate sucrose extraction percentage for the 3rd mill in a 14-roller milling train?________

   A. 70.0%.   B. 12.5%.

   C. 7.0%.   D. 4.5%.

3. What is the primary use of bagasse from the last mill?________

   A. Straining the juice through a metal sheet.

   B. Feeding the cane to the mills.

   C. Fuel for steam generation.

   D. Concentrating the juice.

Task 3: Blank filling

根据短文内容，把下列句子补充完整，每空填写一个单词。

1. The cane is cut into short lengths by rapidly revolving knives and then passed to a ________ train.
2. In compound ________ , water is applied to the crushed cane and the diluted juices are returned to the mills.
3. The approximate percentage of sucrose in the cane extracted by the 3rd mill is about ________ .
4. The bagasse from the last mill is mainly composed of cane fiber and is primarily used as a fuel for raising ________ .
5. The juice is strained through a ________ metal sheet before being sent to the boiling house.

Task 4: Reading comprehension

根据短文内容，回答下列问题。

1. What is compound imbibition?

____

2. What is the main purpose of the boiling workshop in the juice extraction process?

____

Key（Unit 3 Section 5）

Task 1:

Task 2:

1. B 2. C 3. C

Task 3:

1. milling 2. imbibition 3. 4.5% 4. steam 5. perforated

Task 4:

1. Extracting juice by passing water counter-current to the sliced cane.
2. Clarifying and concentrating the juice.

译文

第五节：蔗汁提取

甘蔗首先用快速旋转的刀具将甘蔗切成小段，然后将其送至压榨机组，例如，压榨机组由两辊式压榨机碾蔗机和 4 个三辊式压榨机组成，总共有 14 个辊，在这些辊中，蔗汁通过压力挤出（图 3–1）。三个轧机中的轧辊处于每英尺轧辊长度 75 ~ 80 吨的压力下，长度范围为 2.5 ~ 7.0 英尺；轧碎能力随辊的长度而变化，生产能力从小尺寸的 500 吨到大尺寸的 5 000 吨每 24 小时（图 3–2）。

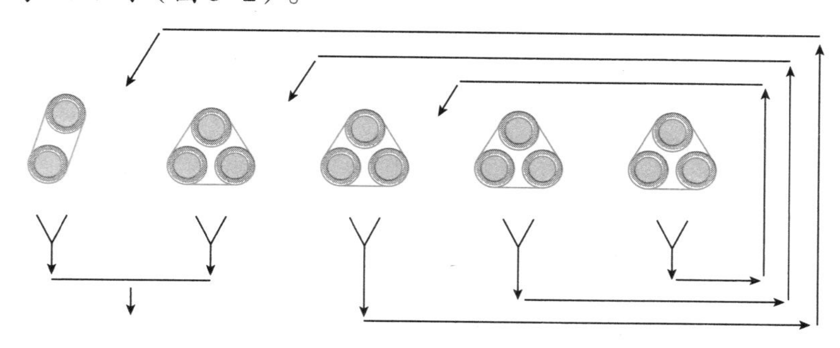

图 3–1　由一台破碎机和四台三辊轧机组成的铣削机组，用水配合碾压；复合渗浸

在碾磨过程中，将水喷淋到破碎好的蔗料上，稀汁返回压榨机。在 14 辊压榨机组中，第三座压榨机处将水喷淋到甘蔗上，并将甘蔗汁供给第一个压榨机；将来自第四座压榨机的稀汁供给至第二座压榨机。这种方法被称为复合渗浸。由压榨机组的每个单元提取的甘蔗中蔗糖的大致百分比见表 3–2：

表 3–2　压榨机组的各单元提取的蔗糖分

蔗刀机	1压榨机	2压榨机	3压榨机	4压榨机	综合
70.0	12.5	7.0	4.5	2.0	96.0

作为提高蔗糖回收率的一种手段，目前正在研究通过水逆流到蔗料碎片中提取甘蔗汁。

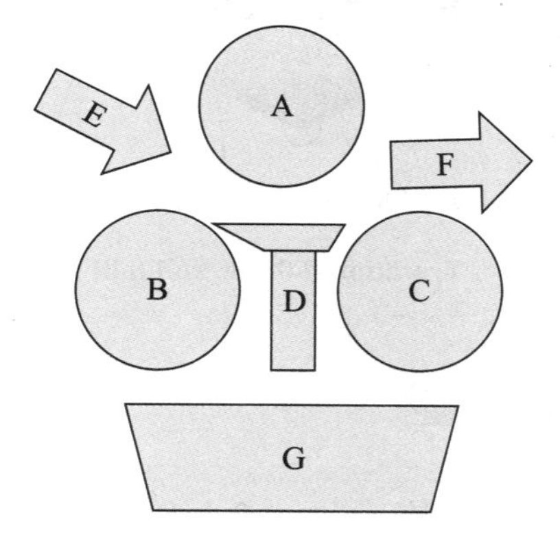

图 3–2　一种甘蔗压榨机有三个压榨辊和一个典型的转盘装置。液压活塞保持 A 上的压力，而 B 和 C 是固定的；轧辊处于每英尺轧辊宽度 75 ～ 80 吨的压力下

A— 顶辊；B— 后辊；C— 进料辊；D— 底梳，甘蔗的进料口；E— 排出蔗渣；F— 集汁槽

蔗汁的成分因土壤和气候条件、甘蔗品种和成熟度而变，在表 3–3 规定的限度内。

表 3–3　蔗汁成分变化区间

水	蔗糖分	还原糖	水不溶物	灰	pH
78%～86%	10%～20%	0.5%～2.5%	0.5%～1.0%	0.3%～0.7%	5.1～5.7

如pH值所示，蔗汁是酸性的；它通过穿孔的金属片过滤，并被送到制炼车间，在那里通过煮沸进行澄清和浓缩。来自最后的压榨机的蔗渣含有52%～58%的固溶物，主要是甘蔗纤维；剩余部分是水。主要用作工厂所需蒸汽的燃料。

Section 6　Identify the picture

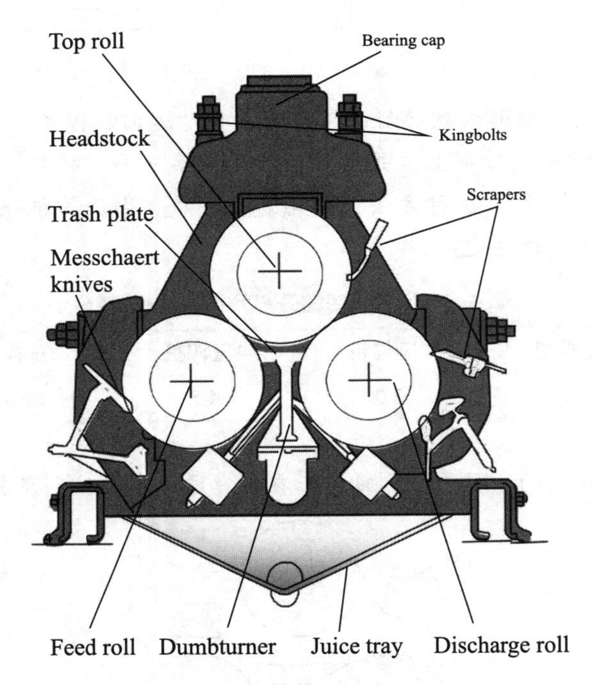

Traditional three-roll mill

New words and expressions

1. scraper /'skreɪpər/ 面梳、后梳
2. kingbolt /'kɪŋˌboʊlt/ 主螺栓
3. traditional three-roll mill 典型三辊压榨机
4. top roll 顶辊
5. bearing cap 轴承盖
6. headstock 主轴箱
7. trash plate 底梳
8. messchaert knives 疏汁刀
9. feed roll 前辊（进料辊）
10. dumbturner 梳桥
11. juice tray 蔗汁盘、蔗汁槽
12. discharge roll 后辊（蔗渣辊）

Exercises

Task1: Fill in the blanks

认真观察图片，并在空白处填上正确的单词

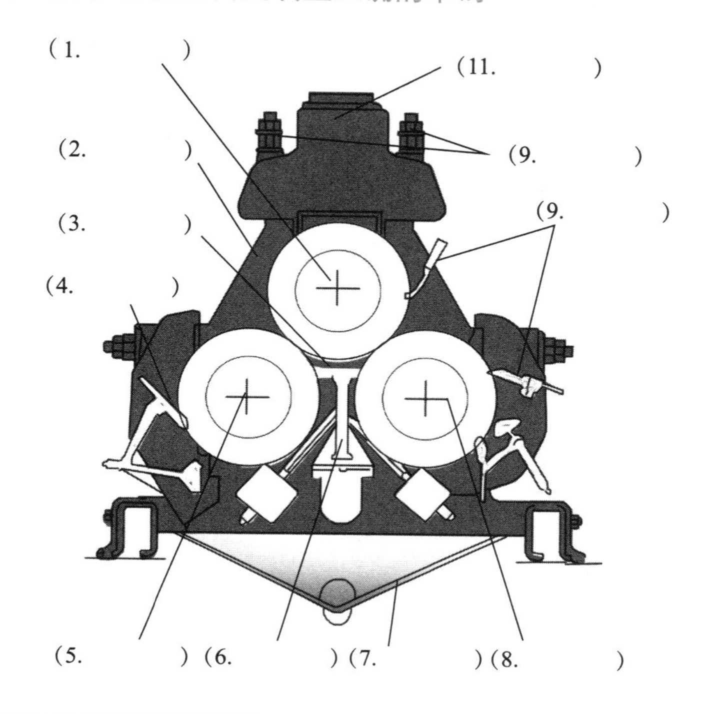

Key（Unit3 Section6）

Task1:

1.top roll 顶辊	2.headstock 主轴箱
3.trash plate 底梳	4.messchaert knives 疏汁刀
5.feed roll 前辊（进料辊）	6.dumbturner 梳桥
7.juice tray 蔗汁盘、蔗汁槽	8.discharge roll 后辊（蔗渣辊）
9.scraper 面梳、后梳	10.kingbolt 主螺栓
11.scraper /ˈskreɪpər/ 面梳、后梳	12.bearing cap 轴承盖

Culture Background

Mill of Ancient China

China and India are the first countries in the world to design and use a double-roller press. As early as the 9th century, there was a description of the machine used to press sugarcane in Dunhuang, China. In the 17th century, "Tian Gong Kai Wu" appeared diagram of the vertical double roller press, many scholars believe that this technological innovation is China's contribution to the world's sugarcane pressing technology.

古代中国压榨机

中国和印度是世界上最早设计和使用双辊压榨机的国家。早在 9 世纪，在中国敦煌就有压榨甘蔗所使用的机器的描述。17 世纪时候，《天工开物》就出现了垂直双辊压榨机的图示，不少学者认为，该技术革新是中国对世界甘蔗压榨技术的贡献。

Unit 4 Clarification

Objectives

After learning this unit, you will be able to

1. Expand your vocabulary about the clarification;
2. Understand the simple lime defecation;
3. Understand the method of clarification,e.g. sulphitation process for white sugar.

Section 1 Discussing the process of clarifying sugarcane juice

Characters:
Mr. Sweet—a sugar industry professionals （制糖行业的专业人士）
Tang Qianxi, Lin Huahua, Zhang Jiejing—students majoring in Sugar Engineering and being enthusiastic about sugar crops（糖业工程专业学生，对糖料作物有浓厚兴趣）

Mr. Sweet: Good morning, everyone. Today, we'll be discussing the process of clarifying sugarcane juice. This is an essential（必要的） step in the production of sugar. Now, let's have a conversation to learn more about this topic. Shall we begin?

Tang Qianxi（Tang）: Sure, I'll start. So, what exactly is the purpose of clarifying sugarcane juice?

Mr. Sweet: Great question! Clarification is done to remove impurities and solids from the juice, making it clear and ready for further processing. Now, Lin Huahua, could you explain how it is typically done?

Lin Huahua (Lin): Certainly. The most common method is the addition of a clarifying agent (澄清剂) such as lime or milk of lime (石灰乳). This raises the **pH level**[1] of the juice and forms a precipitate that traps the impurities. After settling, the clear juice is separated from the **sediment**[2].

Zhang Jiejing (Zhang): I see. Is there any other method apart from using clarifying agents?

Mr. Sweet: Yes, indeed. Another method is the use of **enzymatic**[3] clarification, where specific enzymes (酶) are added to break down the impurities into smaller particles. These particles are then easily separated from the juice.

Tang: How do they ensure that the juice is completely clarified?

Mr. Sweet: Good question. To ensure complete clarification, the juice is often passed through a series of **filter beds**[4] or **centrifugal**[5] separators (离心分离机) to remove any remaining impurities.

Lin: That makes sense. And what happens to the impurities that are removed?

Mr. Sweet: Excellent question. The impurities, also known as filter cake or mud, are typically used as **fertilizer**[6] or processed into **biogas**[7].

Zhang: Thank you for the explanation. It's fascinating to learn about the process of clarifying sugarcane juice.

Mr. Sweet: You're welcome. It's important to understand these processes as it helps us appreciate (欣赏) the **meticulous**[8] steps involved in producing such everyday **commodities**[9]. Keep exploring and asking questions, everyone!

音频：Unit 4 Section 1

New words and expressions

1. pH level		酸碱值；酸碱度
2. sediment	/ˈsedɪmənt/	*n.* 沉积物；沉淀物
3. enzymatic	/ˌɛnzaɪˈmætɪk/	*adj.* 酶促的；酶催的
4. filter beds		过滤床；滤水池
5. centrifugal	/ˌsentrɪˈfjuːgl/	*adj.* 离心的
6. fertilizer	/ˈfɜːtəlaɪzə(r)/	*n.* 肥料
7. biogas	/ˈbaɪəʊgæs/	*n.* 沼气
8. meticulous	/məˈtɪkjələs/	*adj.* 细心的
9. commodity	/kəˈmɒdəti/	*n.* 商品；有用的东西

Exercises

Task 1: Crosswords puzzle

请根据中文提示把下面字谜中的单词填写出来。

Across（横排）

3. 分离
5. 杂质
6. 沼气
8. 固体
10. 酶促的
11. 澄清

Down（竖排）

1. 细心的
2. 肥料
4. 商品
7. 沉淀物
9. 处理

Task 2: One-choice question

根据对话内容，选择下列问题的正确答案。

1. What is the purpose of clarifying sugarcane juice?________

   A. To add impurities to the juice. B. To remove impurities from the juice.

   C. To change the color of the juice. D. To increase the sweetness of the juice.

2. What is the most common method of clarifying sugarcane juice?________

   A. Enzymatic clarification. B. Passing through filter beds.

   C. Adding milk of lime. D. Increasing the pH level.

3. What happens to the impurities removed from the juice?________

A. They are used as fuel.
B. They are used as fertilizer.
C. They are converted into biogas.
D. They are added back to the juice.

Task 3: Blank filling

根据对话内容，把下列句子补充完整，每空填写一个单词。

1. Clarification is done to remove ________ and solids from the juice.
2. The impurities removed from the juice are also known as ________.
3. The addition of ________ raises the pH level of the juice during clarification.

Task 4: Thinking training

根据对话内容，绘制澄清甘蔗汁的思维导图并复述。

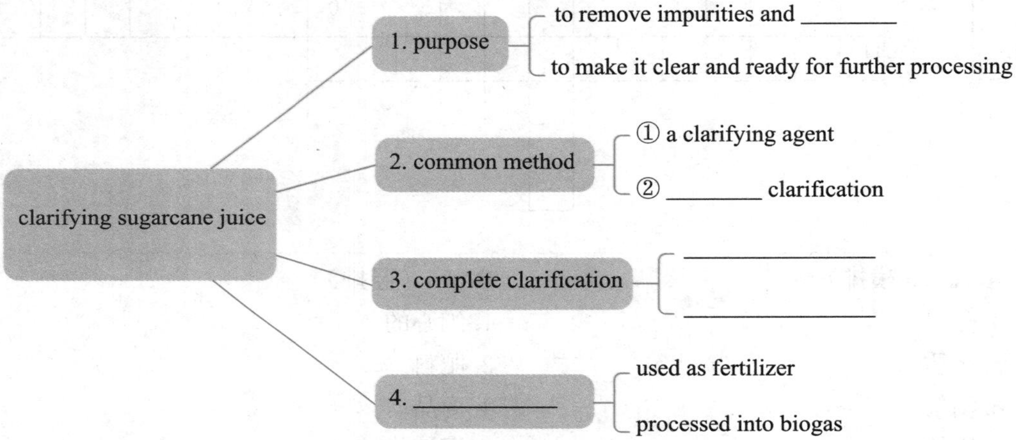

clarifying sugarcane juice

1. purpose
   - to remove impurities and ________
   - to make it clear and ready for further processing
2. common method
   - ① a clarifying agent
   - ② ________ clarification
3. complete clarification
   - ________
   - ________
4. ________
   - used as fertilizer
   - processed into biogas

____
____
____
____
____
____
____
____
____
____
____
____

Key (Unit 4 Section 1)

Task 1:

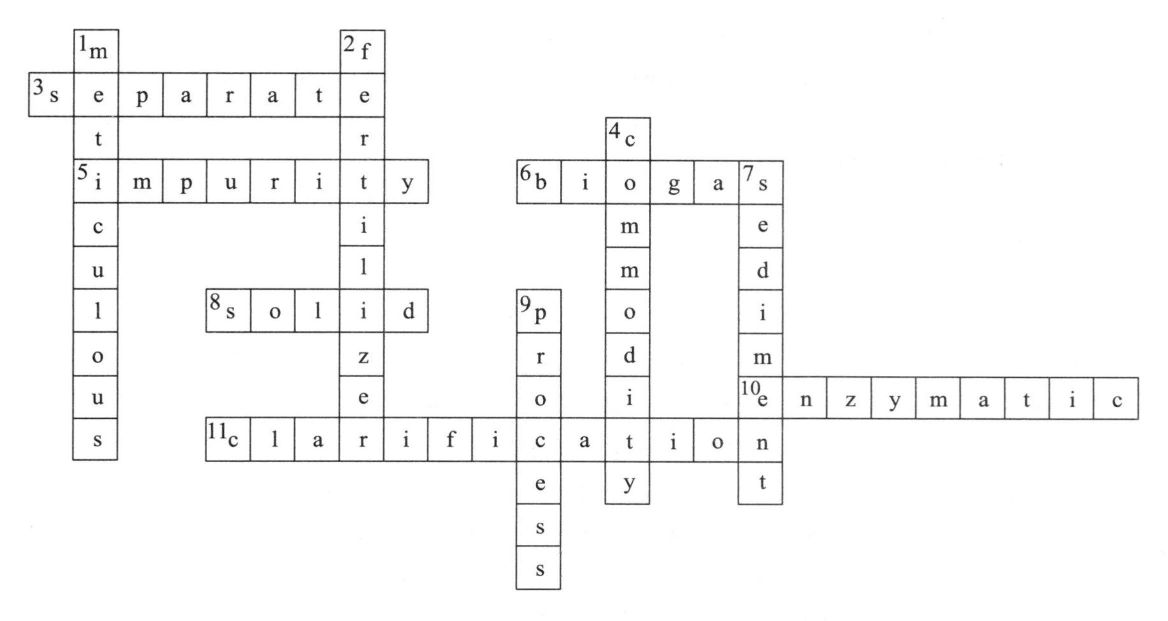

Task 2:

1. B 2. C 3. B

Task 3:

1. impurities 2. filter cake or mud 3. milk of lime

Task 4:

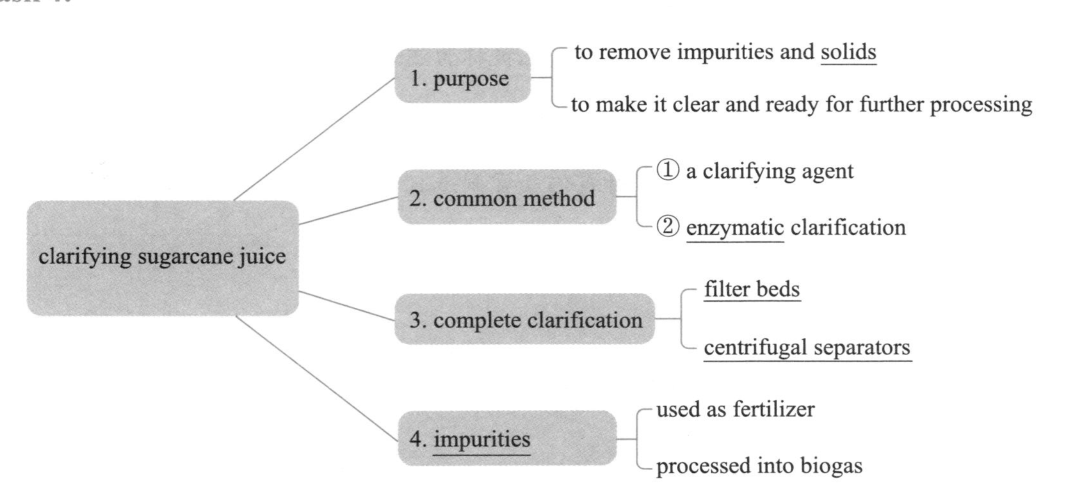

The process of clarifying sugarcane juice is as follows: The purpose is to remove impurities and solids. A clarifying agent such as lime or milk of lime is added to raise the pH level and form a precipitate to trap impurities. After settling, the clear juice is separated from the sediment. Another method is enzymatic clarification where specific enzymes are used to break down impurities into smaller particles. To ensure complete clarification, the juice is passed through a series of filter beds or centrifugal separators. The impurities removed, also known as filter cake or mud, are usually used as fertilizer or processed into biogas.

译文

第一节：讨论澄清甘蔗汁的过程

甜先生：早上好，大家。今天我们将讨论澄清甘蔗汁的过程。这是制糖过程中的一个重要步骤。现在，让我们通过对话来更深入地了解这个话题。我们开始吧？

唐千禧（唐）：当然，我来开始。那么，澄清甘蔗汁的目的究竟是什么？

甜先生：很好的问题！澄清的目的是将汁液中的杂质和固体物去除，使其清澈并且可以进一步处理。现在，林华华，你可以解释一下通常如何进行吗？

林华华（林）：当然。最常用的方法是添加澄清剂，比如石灰或石灰乳。这会提高汁液的 pH 值，形成沉淀物来捕捉杂质。沉淀后，清澈的汁液与沉淀物分离。

张结晶（张）：我明白了。除了使用澄清剂之外，还有其他方法吗？

甜先生：是的，还有一个方法是使用酶澄清，添加特定的酶用于将杂质分解成更小的颗粒。然后这些颗粒很容易与汁液分离。

唐：那么它们是如何确保汁液完全澄清的呢？

甜先生：很好的问题。为了确保完全澄清，通常会将汁液通过一系列的过滤床或离心分离机来去除所有剩余的杂质。

林：那很合理。被去除的杂质会发生什么呢？

甜先生：很棒的问题。被去除的杂质，也被称为滤饼或滤泥，通常被用作肥料或加工成沼气。

张：谢谢你的解释。学习澄清甘蔗汁的过程真是很有意思。

甜先生：不客气。了解这些过程对我们欣赏制造日常商品所涉及的细致步骤很重要。大家继续探索和提问吧！

Section 2 Defecation

The juice from the strainer is clouded by small amounts of **suspended**[2] and colloidal matter, and hence must be **clarified**[3]. For the manufacture of raw cane sugar simple lime **defecation**[1] is practiced. Milk of lime is added to the juice in an amounts sufficient to raise the pH to 7.6 or 1.8. The limed juice is pumped through high-velocity juice heaters（高速蔗汁加热器）(heat exchangers, steam versus juice); insoluble calcium phosphates（不溶性磷酸钙）, sulfates and other salts absorb and entrain in their precipitation a large proportion of colloidal impurities (waxes, pectins and pentosans), the **flocculation**[4] of which is also promoted by the alkaline reaction（碱性反应）. In addition, insoluble proteinates are formed, and the elimination（消除） of proteins is fairly complete. Should the juice contain less than 0.03 of a gram of P_2O_5, per 100 milliliters, enough phosphoric acid（磷酸盐） or superphosphate（复合磷肥） is added to bring the amount up to that figure. Precipitation may take place in setting tanks（设置槽） or in continuous shelf clarifiers（连续式层流澄清器）. The juice emerges from these in crystal clear condition and is pumped to the evaporators. The sediment from the settlers is filtered and the filtrate returned to the subsiders; the solids are sent to the fields as fertilizer.

音频：Unit 4　Section 2

New words and expressions

1.	defecation	/ˌdefəˈkeɪʃn/	*n.* 澄清；净化
2.	suspend	/səˈspend/	*v.* 悬浮，漂浮
3.	clarified	/ˈklærəfaɪd/	*adj.* 澄清的；透明的
4.	flocculation	/flɒkjʊˈleɪʃən/	*n.* 絮凝；絮结产物

Exercises

Task 1: Crosswords puzzle

请根据中文提示把下面字谜中的单词填写出来。

Across（横排）	Down（竖排）
2. 沉淀物	1. 肥料
5. 絮凝	3. 透明的
7. 沉淀	4. 悬浮
8. 澄清	6. 胶质的

Task 2: One-choice question

根据文章内容，选择下列问题的正确答案。

1. Why is lime defecation practiced in the manufacture of raw cane sugar?________
   A. To raise the pH level of the juice. B. To clearing colloidal matter in the juice.
   C. To eliminate proteins from the juice. D. To promote flocculation of impurities.
2. What is the purpose of adding phosphoric acid or superphosphate to the juice?________
   A. To clarify the juice. B. To raise the pH level of the juice.
   C. To promote flocculation of impurities. D. To eliminate proteins from the juice.
3. How is the sediment from the setters treated?________
   A. It is filtered and returned to the subsiders. B. It is sent to the fields as fertilizer.
   C. It is pumped to the evaporators. D. It is mixed with lime for defecation.

Task 3: Blank filling

根据文章内容，把下列句子补充完整，每空填写一个单词。

1. The juice is ________ by small amounts of suspended and colloidal matter.
2. Lime defecation is used to ________ the juice and promote the precipitation of impurities.
3. The filtrate from the sediment is ________ to the subsiders.

Task 4: Thinking training

根据文章，将下列描述重新排序，并试着复述石灰澄清法的步骤。

A. The limed juice is pumped through high-velocity juice heaters.
B. The juice emerges from these in crystal clear condition and is pumped to the evaporators.
C. Adding enough phosphoric acid if the juice contain less than 0.03 of a gram of P_2O_5, per 100 milliliters.
D. The juice from the strainer is clouded by small amounts of suspended and colloidal matter.
E. Precipitation takes place in setting tanks.
F. A large proportion of colloidal impurities are absorbed and entrained in the precipitation.
G. The sediment from the settlers is filtered and the filtrate returned to the subsiders; the solids are sent to the fields as fertilizer.
H. Adding the milk of lime to the juice in an amount sufficient to raise the pH to 7.6 or 1.8.

1. ________ 2. ________ 3. ________ 4. ________ 5. ________ 6. ________
7. ________ 8. ________

Key (Unit 4 Section 2)

Task 1:

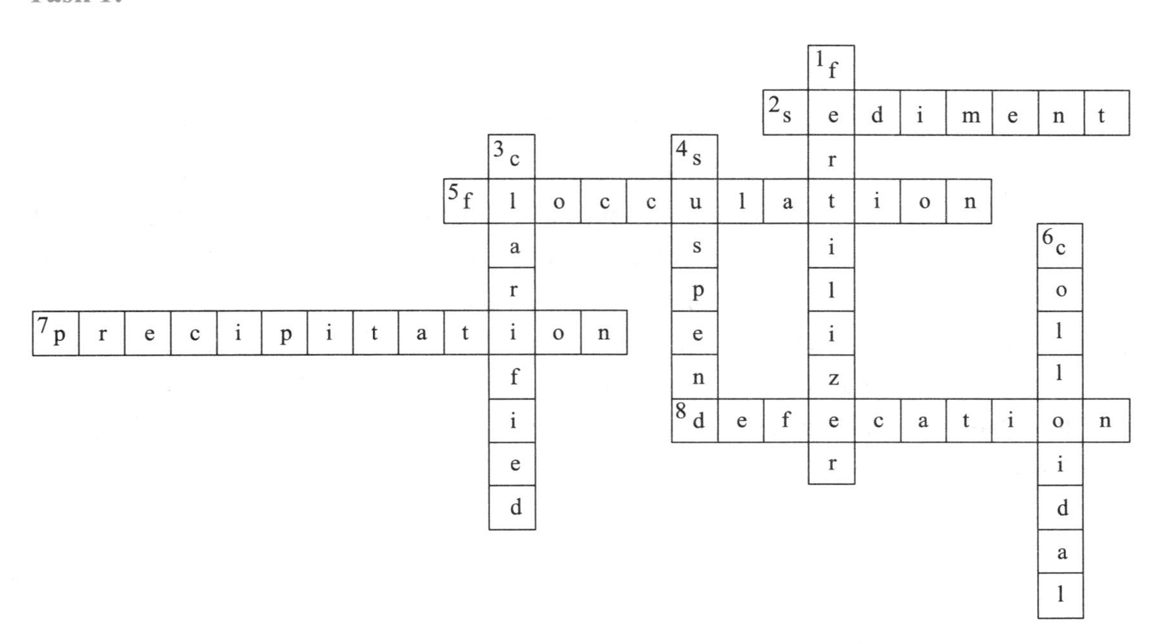

Task 2:

1. B 2. B 3. B

Task 3:

1. clouded 2. clarify 3. returned

Task 4:

1. D 2. H 3. A 4.F 5. C 6. E 7. B 8. G

The steps of lime defecation are as follows:

1. Add milk of lime to the juice in an amount sufficient to raise the pH to 7.6 or 1.8.

2. Pump the limed juice through high-velocity juice heaters.

3. Insoluble calcium phosphates, sulfates and other salts absorb and entrain a large proportion of colloidal impurities in their precipitation. The alkaline reaction promotes the flocculation of these impurities. Insoluble proteinates are formed and the elimination of proteins is fairly complete.

4. If the juice contains less than 0.03 grams of P_2O_5 per 100 milliliters, add phosphoric acid or superphosphate to bring the amount up to that figure.

5. Precipitation may take place in setting tanks or in continuous shelf clarifiers.
6. The juice emerges from these in crystal clear condition and is pumped to the evaporators.
7. The sediment from the settlers is filtered and the filtrate is returned to the subsiders.
8. The solids are sent to the fields as fertilizer.

译文

第二节：澄清

过滤器中的汁液因少量的悬浮物和胶体物质变得混浊，因此需要进行澄清处理。在原蔗糖的制造中采用简单的石灰澄清方法。将石灰乳加入汁液中，加入的量足以将pH提高到7.6或1.8。掺石灰的汁液通过高速蔗汁加热器（热交换器，蒸汽与蔗汁）泵送；不溶性磷酸钙、硫酸盐和其他盐类在它们沉淀过程中吸附并带走大量胶体杂质（蜡、果胶和戊糖等），这种沉淀过程也由碱性反应的促进。此外，还形成了不溶性蛋白酸盐，蛋白质的消除相当彻底。如果汁液中的磷酸盐含量少于每100毫升0.03克P_2O_5，则需要加入足够的磷酸或复合磷肥使其达到该数值。沉淀过程可以在设置槽或连续式层流澄清器中进行。汁液从中以完全澄清的状态出来，然后被泵送到蒸发器。沉淀物经过滤后，滤液返回给沉淀槽，而固体部分作为肥料送往田地。

Section 3 Clarification

The juices extracted from the cane or beet are filter to remove most of the particles of dirt, fiber, or **pulp**[1], after which the juice is ready for the next step, clarification. The purpose of clarification is to free the juice as far as possible from all **constituents**[2] except sugar without altering the sugar itself.

Many materials have been proposed for purification. However, lime, which was one of the first chemicals to be used, still remains the universal basis of most **schemes**[3] of clarification, since it is both an effective and inexpensive material.

The action of lime upon the constituents of the juice is complex and is not completely understood. The lime **neutralizes**[4] the natural **acidity**[5] of the juice and converts many of the **organic**[6] acids into **insoluble**[7] calcium salts. The change in the hydrogenion concentration（氢离子浓度） produced by the alkalinity of the lime prevents any **inversion**[8] of sucrose, and is also responsible for the precipitation of many colloidal impurities. Further the various precipitates that are formed will adsorb still other impurities, and will also trap much of the finely divided(分裂的) suspended matter.

Expressed numerically(数字上), the clarification does not **accomplish**[9] any great removal(移除) of impurities. However, the clarification does produce a remarkable change in the **physical**[10] character（特性）of the juice. The **appearance**[11] which was dark and murky（混浊的）, after clarification, becomes clear and lighter in color. The improved physical character is definitely reflected（显示） in an improved performance in many of the subsequent（随后的）operations, including **filtration**[12], evaporation and in the recovery of sugar in the crystallization（结晶）. Because of all this, clarification remains an essential and integral(必需的）part of the **manufacture**[13] of sugar.

Methods[14] of Clarification:

There are many methods of applying lime for clarification. A simple method (Fig.4-1) of clarifying or defecating cane juice is to add sufficient milk of lime to produce a neutral reaction. This heating is necessary to promote（促进） the clarifying action of the lime, and is also essential to check all biological(生物的）processes which would otherwise rapidly occur(发生）in freshly clarified juices and cause deterioration. Defecation with lime may be carried out as a batch or a **continuous**[15] operation.

Fig.4-1 Flow sheet of simple clarification

Batch System:

In a typical batch system, the juice is placed in tanks to which the required amount of lime is added. The limed juice is then heated until a heavy blanket of scum forms, after which the precipitated matter is allowed to subside or rise, according to its specific gravity(比重). The clear juice is decanted between the scum and settled mud, and placed in a storage tank（储罐）. The mud and scums are sent to a press and the clear filtrates(压滤机）are added to the decanted liquor from the feeding tanks（送料槽）. The juice is then ready for the evaporation station（蒸发站）.

Continuous Process:

In a continuous process the juices flow(流动)continuously through a treatment tank to which the regular addition of the proper quantity（定量） of lime is controlled by a pH meter（pH 计）. From the treatment tank the limed juices flow through a clarifier of the Dorr type(多尔型澄清器), which separates the muds from the juices, and the latter go to the evaporator station.

Some cane juices are rather refractory（难以控制的） to the simple defecation treatment just described. The mud do not settle or filter well, and the decanted and filtered juices do not have good clanty（净度）. Many of the newer varieties（新品种） of cane which have been developed for disease resistant properties（抗病性） and for higher sugar content, provide juices that are inherently difficult to clarify, and this condition is intensified by the efforts to obtain a greater extraction of the juice which at the same time extracts more of the undesirable non-sugars（非糖类物质）. The problem of refractory juices has been met by various modifications of the simple defecation.

Fractional（分段式） Liming:

One process, known as "fractional liming" (Fig.4-2), is based on the knowledge that there are various types of colloidal impurities present in sugar juice. Some of these colloids are precipitated at one hydrogenion concentration, whereas other colloids are precipitated at a different hydrogenion concentration. Therefore, instead of adding all the lime at one time, it is added in successive(依次的）portions, thereby changing the hydrogenion concentration in a series of steps. This accomplishes the precipitation of the different types of colloids.

CANE JUICE FROM MILL

▼

SUFFICIENT MILK OF LIME ADDED TO ADJUST PH TO APPROXIMATELY 6.3

▼

HEATED TO 212 ℉ to 215 ℉

▼

MORE MILK OF LIME ADDED TO ADJUST PH TO APPROXIMATELY 7.6

▼

SETTLED

Fig. 4-2 Flow sheet of fractional liming

Compound Clarification:

Another procedure, known as "compound clarification", takes advantage of the fact that the juice from the crushers（蔗刀机） and the first one or two mills in the roll mill（压榨机） tandem（串联） is easier to clarify than the juice from the latter mills in the tandem. Therefore, in this procedure the juice from the latter stage is kept separate, and is given an initial defecation before being mixed with the juice from the crusher and the first mill for the regular defecation. The juice extracted by the last part of the roll mill tandem thereby receives a double defecation.

Other Defecation Agents:

Various other agents which elaborate the simple defecation are frequently employed. Sometimes these are adopted because of the refractory nature of the juice, and sometimes because they are considered valuable to the finished quality of the sugar. Two of the more important of these agents are sulfur dioxide（二氧化硫） and phosphoric acid（磷酸）or mono calcium phosphate. In addition to their specific clarifying properties, both sulfur dioxide and phosphoric acid are satisfactorily precipitated by lime, and such precipitates possess an adsorptive（吸附的） power for many of the colloidal impurities.

The Carbonation Process:

In a procedure known as "carbonation"（碳酸化） or "Baochong", an excess amount of lime is added to the juice. The excess lime is then precipitated as calcium carbonate by the addition of CO_2, The application of this process is not general in the case of cane juice, but it is almost universal for beet juice. This is because the simple defecation with lime that is effective for producing raw sugar from cane juice does not crystallization of white sugar from beet juices.

The quantity of lime required in the carbonation process usually amounts to from 1.5% to 2.5% of the weight of the beets. In plants where a Steffens process（斯蒂芬斯工艺） is employed, the saccharate of lime（糖化钙盐） from that process is added to supply all or part of the necessary lime. The carbonation is generally carried out in two stages, and either a batch or continuous system can be used. The continuous process has the advantages generally inherent in a continuous method, and in addition, forms a more granular（颗粒） precipitate that is easier to settle and to filter. There are many variations in the carbonation procedure. The following represents the practice as carried out in one of the newer plants.

（1）Typical Carbonation Process:

The juice coming from the diffusion cells is pumped through heaters which raise the temperature to about 90 ℃, from which the juice goes to a tank where the lime is added. The limed juice then goes through a carbonating tank（饱充罐） where CO_2 is bubbled through the juice till pH reach about 11.0. From here the juice flows through a Dorr Thickener（多尔浓缩器） which separates the juice from the sludge（泥汁）, and the sludge is delivered to drum filters（真空转鼓过滤机）. The clear juice from the clarifier（澄清器）is pumped through a second carbonation tank for further treatment with CO_2, where the pH is reduces to about 9.0. The clear juice from this is treated with SO_2, to adjust the pH to approximately（大约）7.5, again filtered, and is ready for the evaporator station.

Certain precautions（预防措施）should be observed in the carbonation. The first carbonation must not be carried too far as otherwise some of the precipitated impurities will be redissolved（再溶解）, and in the second carbonation it is necessary to avoid carbonation to the point where bicarbonates（碳酸氢盐） will form, sines these will be decomposed later in the evaporator, and cause incrustation. Formerly the end point(终点）of carbonation was determined by the appearance of precipitate, and titration（滴定）. Today conductivity and pH instruments（电导率和 pH 仪） are generally employed for this purpose. Some installations are arranged to provide a utomatic control of the CO_2 valves（自动控制二氧化碳阀门）. it is necessary to provide good contact between CO_2, and the liquid phase, and this is assisted by introducing CO_2, in the form of very fine bubbles（气泡）.

（2）Use of Active Carbon in Clarification:

Many beet sugar factories add active carbon（活性炭） at the clarification station（澄清站）. Dosages（计量） as low as one quarter pound（1/4 磅） active carbon per ton of beets have been found to result in improved behavior of the liquor during the subsequent evaporation and crystallization.

Sulphitation Process[16] for White Sugar

A modification(改良)of the sulphitation process that involves a minimum of sucrose inversion (蔗糖转化) and produces a good plantation or near-white sugar in conducted as follows: The cold raw juice is heavily overlimed, to very strong alkalinity (强碱性). The lime is **saturated**[17] with **sulphurous acid**[18], obtained by burning sulphur in a special stove. The sulphitation is continued until the juice remains slightly acid to **phenolphthalein**[19], and is heated to near 90 ℃, to break up lime salts(石灰盐). A **voluminous**[20] precipitate is formed, which is removed by settling, or by filter-pressing (压榨过滤) if the use of lime has been very large. The clear juice must be slightly acid **so as to**[21] hold iron salts in solution. The clarified juice is concentrated to a syrup of about 54 °Bx or higher, which is cooled and sulphited to **distinct**[22] acidity to phenolphthalein. This sulphitation is followed by heating to 90 ℃, settling and decantating. The clear syrup is now ready to boil for sugar.

The Batch process (间歇法), extensively used in Java (爪哇), has recently been improved and is now conducted as follows proportion to total amount: 30% to 40% milk of lime of 26.5 °Bx (锤度) is added to cold raw juice and then sulphited to **faint**[23] acidity to phenolphthalein. The sulphited juice is heated to full boiling(完全沸腾), then settled, and the clear juice is decanted and evaporated to a syrup of 55 to 60 ° Bx. The syrup as it flows from the multiple effect evaporator (多效蒸发器) without heating or cooling, is sulphited, and then 3 to 4 parts of milk of lime (26.5 °Bx) is added and the sulphitation is continued to slight acidity to phenolphthalein. The sulphitation is followed by heating the syrup to 80 to 90℃, and filtration through **cloth**[24] in presses. It is **claimed**[25] that this large use of lime in the syrup promotes the elimination of lime salts more efficiently than a smaller quantity and thus reduces the scaling(结垢)of the heating surfaces of the vacuum pans(真空罐). The **moderate**[26] use of lime in the juice results in little scaling of the juice-heaters(加热器) and evaporators. The syrup filtration is rapid, from 6 to 8 gals (加仑) per hour per square foot (平方英尺) of cloth.

There is a great tendency (趋势) in all sulphitation processes for the heating surfaces in contact with juice or syrup to scale. An acid reaction is essential to keep sulphites in the soluble (可溶的) form, and iron, which is usually present, in **ferrous**[27] state.

音频：Unit 4　Section 3

New words and expressions

1. pulp	/pʌlp/	*n.* 果肉
2. constituent	/kən'stɪtʃuənt/	*n.* 成分
3. scheme	/skiːm/	*n.* 方案
4. neutralize	/'njuːtrəlaɪz/	*v.* 中和
5. acidity	/ə'sɪdəti/	*n.* 酸度
6. organic	/ɔr'gænɪk/	*adj.* 有机的
7. insoluble	/ɪn'sɑləbl/	*adj.* 难溶解的
8. inversion	/ɪn'vɜːrʒən/	*n.* 转化

9. accomplish	/ə'kʌmplɪʃ/	*v.* 完成
10. physical	/'fɪzɪkl/	*adj.* 物理的
11. appearance	/ə'pɪrəns/	*n.* 外表，外观
12. filtration	/fɪl'treɪʃən/	*n.* 过滤
13. manufacture	/ˌmænju'fæktʃə(r)/	*v.* 制造
14. method	/'meθəd/	*n.* 方法
15. continuous	/kən'tɪnjuəs/	*adj.* 连续的
16. sulphitation process		亚硫酸法
17. saturate	/'sætʃəreɪt/	*v.* 使饱和
18. sulphurous acid		亚硫酸
19. phenolphthalein	/ˌfiːnɒl'(f)θæliːn/	*n.* 酚酞
20. voluminous	/və'luːmɪnəs/	*adj.* 很多的，大量的
21. so as to		如此以便（做……）
22. distinct	/dɪ'stɪŋkt/	*adj.* 清楚的，明显的
23. faint	/feɪnt/	*adj.* 模糊的，微弱的
24. cloth	/klɒθ/	*n.* 滤布
25. claim	/kleɪm/	*v.* 声称
26. moderate	/'mɒdərət/	*adj.* 中等的；适度的
27. ferrous	/'ferəs/	*adj.* ［化学］（含）铁的；亚铁的

Exercises

Task 1: Crosswords puzzle

请根据中文提示把下面字谜中的单词填写出来。

Across（横排）	Down（竖排）
3. 大量的	1. 模糊的
7. 酚酞	2. 中等的
8. 滤布	4. 使饱和
	5. 清楚的
	6. 声称

Task 2: One-choice question

根据文章内容，选择下列问题的正确答案。

1. The modification of the sulphitation process aims to produce ________.
   A. dark brown sugar
   B. near-white sugar
   C. heavily overlimed juice
   D. lime saturated with sulphurous acid
2. The purpose of heating the juice in the sulphitation process is to ________.
   A. break up lime salts
   B. remove iron salts
   C. add phenolphthalein
   D. produce lime precipitate
3. The Batch process is extensively used in ________.
   A. Java
   B. China
   C. Brazil
   D. Australia
4. The large use of lime in the syrup is claimed to promote ________.
   A. the elimination of lime salts
   B. the scaling of heating surfaces
   C. the reduction of juice-heaters and evaporators
   D. the filtration through cloth in presses
5. The acid reaction in sulphitation processes is essential to ________.
   A. keep juice in soluble form
   B. keep sulphites in soluble form
   C. keep iron in soluble form
   D. keep syrup from scaling

Task 3: Blank filling

根据文章内容，把下列句子补充完整，每空填写一个单词。

1. The cold raw juice is heavily ________.
2. The sulphitation process continues until the juice remains slightly acid to ________.
3. The clarified juice is concentrated to a syrup of about ________ °Bx.
4. The sulphited juice is heated to full ________.
5. The syrup filtration is rapid, from 6 to 8 gals per ________ per square foot of cloth.

Task 4: Reading comprehension

根据对话内容，回答下列问题。

1. If the use of lime has been very large, by which means could a voluminous precipitate be removed?

____

2. By the Batch process, how many parts of milk of lime of 26.5 °Bx should be added to cold raw juice?

____

Key (Unit 4 Section 3)

Task 1:

						1 f												
						a		2 m										
	3 v	o	l	u	m	i	n	o	u	4 s				5 d				
						n		d		a				i				
						t		e		t				s				
								r		u				t				
					6 c			a		r				i				
7 p	h	e	n	o	l	p	h	t	h	a	l	e	i	n				
					a			e		t				8 c	l	o	t	h
					i					e				t				
					m													

Task 2:

1.B 2.A 3.A 4.A 5.B

Task 3:

1. overlimed 2. phenolphthalein 3. 54 4. boiling 5. hour

Task 4

1. By filter-pressing. 2. 30% to 40%.

译文

第三节：澄清

从甘蔗或甜菜中提取的汁液经过过滤除去大部分的泥土、纤维或果肉颗粒，之后这些汁液便会进入下一个步骤，即澄清。澄清的目的是尽可能地使汁液除了糖之外不含其他成分，并且不改变糖分本身。

为了达到澄清目标，需要将许多物质提取出来。然而，石灰作为早被使用的化学物质之一，仍然是大多数澄清方案的基础，因为它既有效又廉价。

石灰对甘蔗汁成分的作用非常复杂，目前还没有完全解释。石灰中和了甘蔗汁的天然酸度，并将许多有机酸转化为难溶解的钙盐。石灰碱性产生的氢离子浓度的变化防止了蔗糖的转化，并且也沉淀了许多胶体杂质。形成的各种沉淀物将吸附其他杂质，并且会捕捉到大部分细碎的悬浮物质。

从数值上来说，澄清并没有大幅度地去除杂质。然而，澄清确实使甘蔗汁的物理特性产生了显著的变化。在澄清之前，汁液呈现出深色和混浊，在澄清后变得清澈且颜色较浅。改善后的物理特性会在后续的许多操作中，包括过滤、蒸发和结晶回收糖中表现出更好的效果。因此，澄清仍然是生产蔗糖的基本和必要的一部分。

澄清方法：

有许多使用石灰进行澄清的方法。一种简单的方法是向甘蔗汁中加入足够的石灰水，使其呈现中和反应（图 4-1），以澄清或脱色甘蔗汁。加热是必要的，可以促进石灰的澄清作用，还可以基本检查所有在新鲜澄清的汁液中迅速发生的生物过程及腐败。使用石灰进行澄清可以作为批量间接处理或连续操作进行。

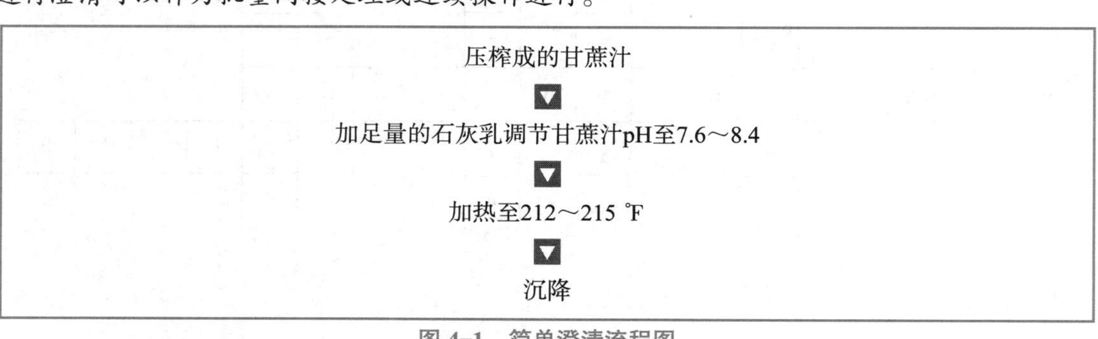

图 4-1　简单澄清流程图

间歇处理系统：

在典型的批量处理系统中，将汁液放入槽中，并加入所需的石灰。然后加热汁液，直到形成大量浮渣，之后根据其比重，沉淀物会下降或上浮。清澈的汁液从渣和沉淀泥之间倾倒出来，并放入储罐中。滤泥和浮渣被送入压滤机，并加入从进料槽中倾倒出来的液体。然后甘蔗汁就可以去蒸发站处理了。

连续处理：

在连续加灰处理中，汁液不断地流过加灰槽，石灰定量添加是通过 pH 计控制的。经过加灰槽后，加了石灰的糖汁经过多尔型澄清器，将浮渣与甘蔗汁分离，后者进入蒸发站。

有些甘蔗汁液相对难用刚才描述的简单沉淀处理方法。泥汁沉淀及过滤效果不好，沉降及过滤后的糖汁净度不高。一些新的抗病性和糖分含量更高的甘蔗品种所产生的甘蔗汁本身很难澄清，而且为了获得更高的提取率，同时提取了更多的非糖类物质，对于难以处理的甘蔗汁问题，需要对简单澄清方法进行各种改进。

分段加灰：

被称为“分段加灰”的方法（图 4–2）是基于糖汁中存在着的各种类型的胶体杂质。其中一些胶体在一个氢离子浓度下沉淀，而其他胶体则在不同的氢离子浓度下沉淀。因此，不是一次性添加所有石灰，而是分段添加，在一系列步骤中改变氢离子浓度。这样可以沉淀出不同类型的胶体。

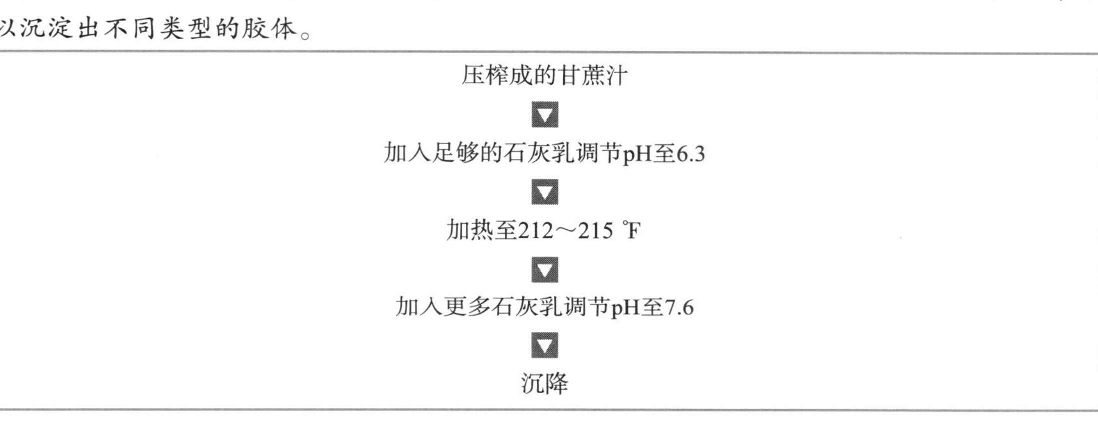

图 4–2　分段加灰流程图

复合澄清：

另一种方法称为“复合澄清”，即蔗刀机和压榨机串联中的前一两个压榨机的甘蔗汁比串联中的最后压榨机得到的甘蔗汁更容易澄清。因此，在这种方法中，后面阶段的汁液被分开处理，在与蔗刀机和第一个压榨机的甘蔗汁混合进行常规澄清之前，先初步沉淀。通过这种方式，压榨机串联的最后部分所提取的蔗汁会进行两次沉淀。

其他澄清剂：

经常使用各种其他澄清剂来对简单澄清进行改进。有时是因为甘蔗汁的抵抗性，有时是因为它们对糖的成品质量有价值。其中两种较为重要的澄清剂是二氧化硫和磷酸或磷酸一钙盐。除了具有特定的澄清性质外，二氧化硫和磷酸都可以被石灰较好地沉淀，而且这种沉淀对许多胶体杂质具有吸附力。

碳酸化工艺：

在一种被称为“碳酸化”或“饱充”的过程中，向甘蔗汁中添加过量的石灰。然后通过添加二氧化碳使多余的石灰沉淀为碳酸钙。甘蔗汁应用碳酸饱充并不普遍，但在甜菜汁液中很普遍。这是因为对于甘蔗汁液，使用石灰的简单澄清可以有效地制备原糖，但无法从甜菜汁液中结晶出白糖。

碳酸饱充过程中所需的石灰量通常相当于甜菜质量的 1.5% ～ 2.5%。在使用斯蒂芬斯工艺的工厂中，从该工艺中获得的糖化钙盐被添加到所需的石灰中。饱充通常分为两个阶段进行，使用间歇或连续系统。连续过程具有连续方法固有的优势，并且形成更易于沉淀和过滤的颗粒状沉淀物。碳酸化过程有很多变化。以下是一个较新工厂中的实践。

（1）典型的饱充过程：

从渗浸室中出来的甘蔗汁被泵送到加热器中，将温度升至约 90 ℃，然后进入一个储槽，加入石灰。接着，加灰后的甘蔗汁经过一个饱充罐，二氧化碳被通入甘蔗汁中，直到达到约 11.0 的 pH。然后，汁液通过一个多尔浓缩器，将汁液与泥汁分离，泥汁被运送到真空转鼓过滤机中。从澄清器中流出的清汁被泵送到第二个饱充罐中，进一步与 CO_2 进行处理，使 pH 降至约 9.0。从中得到的清澈汁液接着通过 SO_2 进行处理，将 pH 调节到约 7.5 左右，再次进行过滤，即可送入蒸发站处理。

在饱充过程中需要注意一些预防措施。第一次饱充不应进行得太彻底，否则一些沉淀的杂质将被重新溶解。而在第二次饱充中，需要避免过饱和而使碳酸氢盐形成，因为这些物质在蒸发器中会分解，导致结垢。过去，饱充的完成是通过观察沉淀物的形成和滴定来确定。如今，用电导率和 pH 仪器来确定。一些设备还设有自动控制二氧化碳阀门，确保二氧化碳能与甘蔗汁良好接触，这可以通过非常细小的气泡形式引入二氧化碳实现。

（2）在澄清过程中使用活性炭：

许多甜菜糖厂在澄清站中加入活性炭。试验证明，每吨甜菜使用 1/4 磅（1 磅 = 0.454 千克）活性炭的剂量可以改善后续的蒸发和结晶过程中糖汁效果。

亚硫酸法制白糖

亚硫酸法工艺的改良版可以实现最小化蔗糖转化，并且可以制取出优质的耕地白糖或接近白糖的产品，操作步骤如下所述。将冷蔗原汁加入大量石灰至过碱性。然后用特制炉子燃烧硫黄，使石灰与硫酸饱和。磺化反应持续进行，直到原汁酚酞指示剂仍然稍微呈酸性，然后加热至约 90 ℃，以分解石灰盐，形成大量的沉淀物，可以沉淀去除，如果使用石灰较多，也可以通过压榨过滤来去除。澄清的汁液必须略呈酸性以便于铁盐的溶解。澄清的汁液浓缩至约 54 °Bx 或更高，然后冷却并亚硫酸化至酚酞指示剂呈酸性，再加热至 90 ℃，沉淀和倾倒。澄清的糖浆就可以开始煮沸制糖。

间歇法是在爪哇广泛使用的工艺，最近进行了改进，操作步骤如下：将占总量 30% ~ 40% 的 26.5 °Bx 石灰乳加入冷的原汁中，亚硫酸化至酚酞指示剂呈微酸性。亚硫酸化的汁液加热至完全沸腾，然后沉淀，将清澈的汁液倾析并蒸发至 55 ~ 60 °Brix 的浓缩糖浆。从多效蒸发器流出的糖浆在不进行加热或冷却的情况下，需要进行亚硫酸化，然后加入 30% ~ 40% 的石灰乳（26.5 °Bx），继续亚硫酸化直至酚酞指示剂呈微酸性。之后，将糖浆加热至 80 ~ 20 ℃，并通过过滤布进行过滤。据称，糖浆中大量使用石灰可以更有效地去除石灰盐，减少真空罐的加热表面结垢。汁液中适度使用石灰产生少量加热器和蒸发器的结垢。糖浆过滤速度快，约为每平方英尺滤布每小时 6 至 8 加仑[1 加仑(美)=3.785 升]。

在所有的亚硫酸法工艺中，与汁液或糖浆接触的加热表面很容易结垢。保持酸性反应对于使亚硫酸盐保持溶解形态是必要的，同时铁离子也能保持在亚铁态。

Section 4 Identify the picture

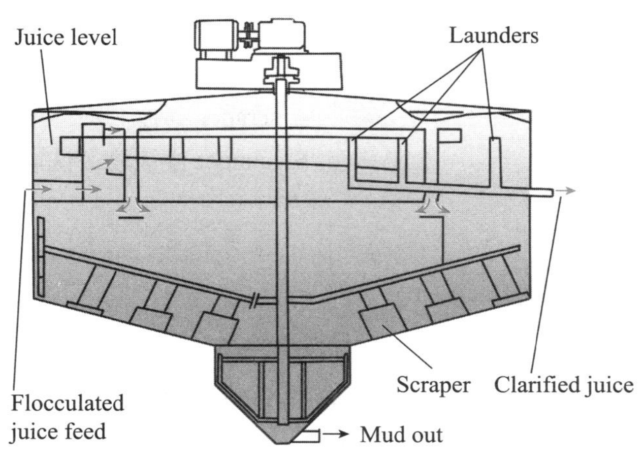

The SRI Rapid Clarifier

New words and expressions

1. launder /'lɔːndər/ *n.* 溢流槽；洗涤口
2. scraper /'skreɪpər/ *n.* 刮板
3. juice level 清汁液面
4. flocculated juice feed 絮凝后蔗汁进口
5. mud out 泥汁出口
6. clarified juice 清汁

Exercises

Task1: Fill in the blanks

认真观察图片，并在空白处填上正确的单词

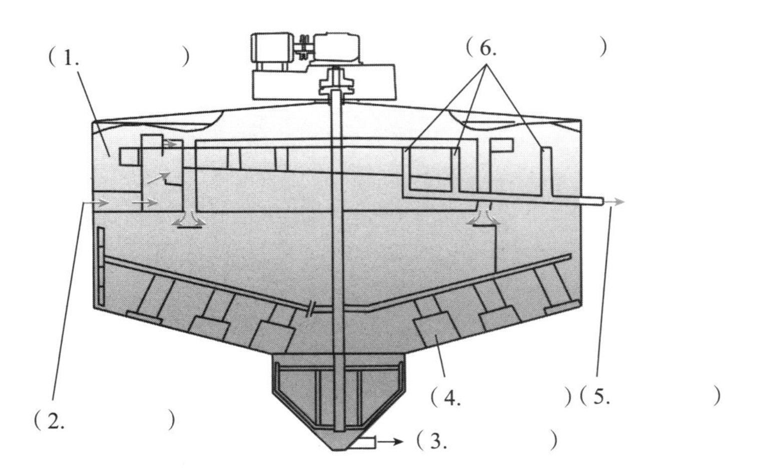

Key（Unit4　Section4）

Task1:

1. juice level 清汁液面
2. flocculated juice feed 絮凝后蔗汁进口
3. mud out 泥汁出口
4. scraper 刮板
5. clarified juice 清汁
6. launder 溢流槽；洗涤口

Culture Background

Yellow Mud Water Spraying Method

After the Song Dynasty, China's sugar production technology stagnated, and the clarification technology of cane juice was not well developed. It was necessary to import craftsmen from Egypt to guide the use of wood ash for clarification. In the 7th century, milk was widely used in India and Persia to clarify cane juice to prepare lighter-colored sugar. In the Yuan Dynasty, China began to use clay filtration technology to produce relatively high-grade white sugar, which greatly reduced production costs. Then, the emergence of the epoch-making "yellow mud water spraying method" enabled China to produce the best white sugar in the world at that time. A large number of exports aboard.

黄泥水淋法

在宋代以后，中国制糖技术有所停滞，甘蔗汁澄清技术没有得到很好的发展，需要从埃及输入工匠来指导使用木灰进行澄清。7 世纪，印度和波斯已经广泛使用牛奶来澄清甘蔗汁以制备颜色较浅的糖。到了元朝，中国开始出现黏土过滤技术以生产较为高级的白糖，生产成本大大降低，随后，具有划时代意义的“黄泥水淋法”的出现，使中国能生产出当时世界上最好的白糖并大量出口国外。

Unit 5 Evaporation

Objectives

After learning this unit, you will be able to

1. Expand your vocabulary about evaporation;
2. Understand the principles and process of evaporation;
3. Develop teamwork spirit and improve interpersonal skills by cooperating with peers.

Section 1 Discussing the principles of evaporation

Characters:
Mr. Sweet—a sugar industry professionals （制糖行业的专业人士）
Zhang Jiejing—a student majoring in Sugar Engineering and being enthusiastic about sugar crops (糖业工程专业学生，对糖料作物有浓厚兴趣)

Mr. Sweet: Good morning, class! Today, we will be discussing the principles of evaporation in a sugarcane factory. Can anyone tell me the purpose of evaporation in this process?

Zhang Jiejing: Is it to remove excess （多余的） water from the cane juice?

Mr. Sweet: Exactly! The primary goal of evaporation is to concentrate the cane juice by removing water, ultimately turning it into syrup（糖浆）. But why do we use **multiple-effect**[1] evaporation in sugar factories?

Zhang Jiejing: I think it's to improve efficiency and save energy.

Mr. Sweet: That's correct! Multiple-effect evaporation helps achieve higher energy **efficiency**[2] by utilizing（利用） the heat generated from one effect to drive the next effect. This way, we can reduce fuel consumption（消耗） and operational costs. Can anyone tell me the key indicators（指标） or measures used in the evaporation process?

Zhang Jiejing: Is it the **Brix**[3] scale to measure the concentration of the syrup?

Mr. Sweet: Well done! The Brix scale is commonly used to measure the sugar concentration in the syrup during the evaporation process. We aim for a specific Brix level to ensure proper crystallization and achieve the desired sugar quality. Additionally, we also **keep a close eye on**[4] evaporation rates and the overall heat transfer efficiency to **optimize**[5] the process.

Zhang Jiejing: Is there anything else we need to consider during evaporation?

Mr. Sweet: Great question! We must also monitor the **scaling**[6] or deposition of impurities on the heating surfaces. Scaling can reduce heat transfer efficiency and increase maintenance costs. Therefore, controlling scaling is crucial to ensure smooth operations. Remember, precise（精确的） monitoring and adjustments are necessary throughout the evaporation process.

Zhang Jiejing: How does the syrup filtration happen after evaporation?

Mr. Sweet: After evaporation, the syrup is typically filtered through cloth in presses to remove any remaining impurities. The filtration rate is determined by factors such as the size of the cloth and pressure applied. This filtration step helps to purify the syrup before it is further processed into granulated（颗粒状的） sugar or other sugar products.

音频：Unit 5　Section 1

New words and expressions

1.	multiple-effect	/ˈmʌltɪplɪfˈekt/	*n.* 多效
2.	efficiency	/ɪˈfɪʃnsi/	*n.* 效率；［物］功率
3.	Brix	/briks/	*adj.* 糖度
4.	keep a close eye on		留心瞧着，注意，密切关注
5.	optimize	/ˈɒptɪmaɪz/	*vt.* 使最优化，使尽可能有效
6.	scaling	/ˈskeɪlɪŋ/	*n.* 结垢

Exercises

Task 1: Crosswords puzzle

请根据中文提示把下面字谜中的单词填写出来。

Across（横排）	**Down（竖排）**
3. 蒸发，消失	1. 效率；[物] 功率
4. 结晶化	2. 使浓缩
6. 过滤；滤除	5. 使最优化
9. 多效	7. 不纯，杂质
10. 糖度	8. 结垢

Task 2: One-choice question

根据对话内容，选择下列问题的正确答案。

1. What is the primary goal of evaporation in a sugarcane factory?________

A. To remove excess water from the cane juice.

B. To increase the Brix level of the syrup.

C. To improve energy efficiency.

D. To achieve higher crystallization.

2. Why do sugar factories use multiple-effect evaporation?________

A. To save fuel consumption.

B. To reduce operational costs.

C. To improve energy efficiency.

D. All of the above.

3. What is the key indicator used to measure the concentration of the syrup?________

A. Evaporation rates.

B. Heat transfer efficiency.

C. Brix scale.

D. Impurity deposition.

Task 3: Blank filling

根据对话内容，把下列句子补充完整，每空填写一个单词。

1. The goal of evaporation is to ________ the cane juice.
2. Multiple-effect evaporation helps achieve higher ________ efficiency.
3. The Brix scale is used to measure the ________ of the syrup.
4. The syrup is typically filtered through ________ in presses.
5. Controlling scaling is crucial to ensure smooth ________.

Task 4: Reading comprehension

根据对话内容，回答下列问题。

1. What is the purpose of syrup filtration after evaporation?

____

____

____

____

____

2. Why is controlling scaling important in the evaporation process?

____

____

____

____

____

Key (Unit 5 Section 1)

Task 1:

Task 2:

1. A 2. D 3. C

Task 3:

1. concentrate 2. energy 3. concentration 4. cloth 5. operations

Task 4:

1. The purpose of syrup filtration after evaporation is to purify the syrup.
2. It improves heat transfer efficiency, reduces maintenance costs and prevents impurity deposition.

译文

第一节：讨论蒸发原理

甜先生：同学们，早上好！今天，我们将讨论甘蔗工厂的蒸发原理。谁能告诉我在这个过程中蒸发的目的是什么？

张结晶：是去除甘蔗汁里多余的水分吗？

甜先生：没错！蒸发的主要目的是通过去除水分来浓缩甘蔗汁，最终将其变成糖浆。但我们为什么要在糖厂使用多效蒸发呢？

张结晶：我觉得是为了提高效率和节约能源。

甜先生：没错！通过利用一种效应产生的热量来驱动下一种效应，多效蒸发有助于实现更高的能源效率。这样，我们可以减少燃料消耗和运营成本。谁能告诉我蒸发过程中使用的关键指标或措施？

张结晶：是用自制糖度来测量糖浆的浓度吗？

甜先生：说得好！自制糖度常用于测量糖浆蒸发过程中的糖浓度。我们瞄准特定的糖度水平，以确保适当的结晶和达到所需的糖质量。另外，我们还密切关注蒸发速率和整体传热效率，以优化工艺。

张结晶：在蒸发过程中还有什么需要考虑的吗？

甜先生：好问题！我们还必须监测受热面上的结垢或杂质沉积。结垢会降低传热效率，增加维护成本。因此，控制结垢对于确保平稳运行至关重要。记住，在整个蒸发过程中都需要精确监测和调整。

张结晶：糖浆蒸发后是如何过滤的？

甜先生：糖浆蒸发后，通常用滤布在压榨机中过滤掉任何残留的杂质。过滤速率是由布的大小和施加的压力等因素决定的。这个过滤步骤有助于在进一步加工成砂糖或其他糖产品之前净化糖浆。

Section 2 Evaporation process

The thin juices from the clarification station must now be evaporated in order to produce crystal sugar. This operation must not be delayed since the thin juice rapidly decomposes(分解) with loss of sugar. Evaporation is conducted by means of steam in evaporators, in the first stage of the evaporation the juice is concentrated to about 50% to 60% sugar. This stage is conducted in multiple effect evaporators which **make** very effective and efficient **use of**[1] the steam.

Multiple Effect Evaporators:

A multiple effect evaporator（多效蒸发器）in the sugar industry consists of a series of large cylindrical（圆柱形的）vessels（容器）(three or four vessels are commonly used, known as triple or quadruple–effect **respectively**[2]). All of the effects in a multiple effect evaporator are of similar construction and design, consisting of a large vertical（垂直的） **cast–iron**[3] or metal plate cylindrical vessel. In the sugar industry the steam–heating unit [usually a calandria（冷却管）] is placed in the lower portion of the effect. The calandria is a drum-shaped steam **chest**[4]（圆筒蒸汽箱）provided with a large vertical central tube [the down take（下管）] which **extends**[5] from the top to the bottom plate of the drum. In the rest of the drum are numerous small-diameter（小直径）vertical tubes also extending from the top to the bottom plate.

Exhaust or low pressure steam is introduced into the calandria of the first effect which also receives the thin juice. The effective heat in the steam is transferred through the walls of the calandria tubes to the juice, causing it to boil. The boiling juice rises in the tubes, releases the vapors（蒸汽） which leave the effect through the **dome**[6] and are piped（管道输送）to the calandria of the **succeeding**[7] effect. The liquor then **descends**[8] at the center of the effect through the "down take" in which no heating takes place. The **effectiveness**[9] of the natural circulation is established depends on proper design of the effect and on the **viscosity**[10] of the liquor. The viscosity increases as the evaporation process, and hence the circulation is somewhat poorer in the last effect.

The vacuum maintained in each of the effects is the force that causes the liquor to flow from one effect to the following one, the rate of flow being controlled by a valve（阀门）. In the case of the last effect（末效罐）, the concentrated liquor is removed by means of a pump, and is ready for final evaporation in the vacuum pans.

Even the clearest clarified juice contain constituents that become insoluble（不溶解的）when the juice is concentrated by water removal. Some of the precipitate（沉淀物）thus formed remains in the thick juices, making it turbid（混浊）, and must be removed by a subsequent filtration. Other impurities are deposited as an **incrustation**[11] on the heating surface of the calandria, particularly in the last effect, such incrustation must be removed from time to time since it slows the transfer of heat and **impairs**[12] the evaporator efficiency.

The vapors leaving the surface of the liquor have a certain velocity（速率）and may carry small **droplets**[13] of liquor a condition known as "**entrainment**"[14]. Entrainment is avoided by proper design of the effect and by keeping the level of the liquor at a correct height. A **baffle**[15] known as a "**catch–all**"[16] is built into the dome of each effect for the purpose of separating any **entrained**[17] droplets from the vapor.

The vacuum is maintained in the last effect by **condensing**[18] the vapors arising from it by means of cooling water. In the other effects the vacuum is obtained by the condensation of the vapor in the calandria of the succeeding effect. The proper degree of vacuum（真空）is controlled by the pump, **trap**[19] or **barometric**[20] drain employed to remove the **condensate**[21] from the calandria and also by a vacuum pump which removes the non–condensable gases.

Crystallization（结晶）:

At this stage the evaporation is continued to the point where sugar crystals form and separate from the remaining water and impurities.

The Role of Supersaturation（超饱和溶液）:

The data on the **solubility**[22] of sucrose（蔗糖）in water at various temperatures are given in Table 5-1.

Table 5-1 Solubility of sucrose in water

Temp/℃	30	40	50	60	70	80
W/*W*/%	68.70	70.42	72.25	74.18	16.22	78.36

If a saturated（饱和的）solution of sucrose is cooled, or if some of the water is evaporated, the solution becomes supersaturated with on **immediate**[23] separation of crystals. The degree of supersaturation is **conveniently**[24] expressed by the **coefficient**[25] of supersaturation, a value obtained（获得）by dividing the amount of sugar in solution by the **theoretical**[26] solubility at that temperature. There is a natural **tendency**[27] for sugar to crystallize from a supersaturated solution, but time is required for this, and the length of time depends on a variety of conditions. When the coefficient（系数）of supersaturation is small, when there is only slight supersaturation, that the sugar will crystallize only on crystals already present in solution. When such crystals are not present, a solution may remain supersaturated for an **indefinite**[28] period of time. However, a solution that is strongly supersaturated, e.g. 1.5 coefficient of supersaturation, will form crystals **spontaneously**[29]. There is an **intermediate**[30] zone of supersaturation, **wherein**[31] new crystals [false grain（伪晶）] will form **in the presence of**[32] crystals already present, but will not form in their absence.

Supersaturation acts as a **positive**[33] force to cause crystallization but there are **negative**[34] influences which **retard**[35] it, one of which is viscosity. A higher coefficient of supersaturation must exist in order to obtain spontaneous crystallization. Sugar factory syrups are not solutions of pure sucrose, but contain non-sugars which not only increase the viscosity, but also, alter the solubility of the sugar. Certain non-sugar has the **apparent**[36] ability to increase the solubility of sugar, where as other non-sugars have the opposite property of decreasing the solubility of sugar. The non-sugar in sugar juices have a very **definite**[37] technical importance in the boiling operation.

音频：Unit 5 Section 2

New words and expressions

1. make... use of		利用
2. respectively	/rɪ'spektɪvli/	*adv.* 分别地

3. cast-iron /ˌkɑːst ˈaɪən/ *n.* 生铁，铸铁
4. chest /tʃest/ *n.* 箱
5. extend /ɪkˈstend/ *v.* 延展，延伸
6. dome /dəʊm/ *n.* 穹顶，圆屋顶
7. succeed /səkˈsiːd/ *v.* 接替，继任
8. descend /dɪˈsend/ *v.* 下降
9. effectiveness /ɪˌfekˈtɪvnɪs/ *n.* 效力
10. viscosity /vɪˈskɒsətɪ/ *n.* 黏度
11. incrustation /ˌɪnkrʌˈsteɪʃn/ *n.* 硬壳，镶嵌，积垢
12. impair /ɪmˈpeə(r)/ *vt.* 损害，减少，削弱
13. droplet /ˈdrɒplət/ *n.* 小水滴，微滴
14. entrainment /ɪnˈtreɪnmənt/ *n.* 雾沫夹带，跑糖
15. baffle /ˈbæfl/ *n.* 挡板
16. catch-all /kætʃ ɔːl/ *n.* 捕汁器
17. entrain /ɪnˈtreɪn/ *v.* 带走，分离
18. condense /kənˈdens/ *v.* 冷凝，浓缩
19. trap /træp/ *n.* 隔栏
20. barometric /ˌbærəˈmetrɪk/ *adj.* 气压
21. condensate /ˈkɒnd(ə)nseɪt/ *n.* 冷凝水
22. solubility /ˌsɒljʊˈbɪlətɪ/ *n.* 溶解度
23. immediate /ɪˈmiːdiət/ *adj.* 直接的，立即的
24. conveniently /kənˈviːnjəntlɪ/ *adv.* 便利地，方便地
25. coefficient /ˌkəʊɪˈfɪʃnt/ *n.* 系数
26. theoretical /ˌθɪəˈretɪkl/ *adj.* 理论的
27. tendency /tendency/ *n.* 倾向，趋势
28. indefinite /ɪnˈdefɪnət/ *adj.* 不确定的，不定的，可能的
29. spontaneously /spɒnˈteɪnɪəslɪ/ *adv.* 自然地，自发地
30. intermediate /ˌɪntəˈmiːdiət/ *adj.* 媒介
31. wherein /weərˈɪn/ *adv.* 在那里，在其中
32. in the presence of 在有……参加的情况下
33. positive /ˈpɒzətɪv/ *adj* 积极的，正向的
34. negative /negətɪv/ *adj.* 否定的，负面的
35. retard /rɪˈtɑːd/ *vt.* 延缓，阻止
36. apparent /əˈpærənt/ *adj.* 明显的
37. definite /ˈdefɪnət/ *adj.* 确定的，无疑的，明确的

Exercises

Task 1: Crosswords puzzle

请根据中文提示把下面字谜中的单词填写出来。

Across（横排）

4. 溶解度
5. 穹顶，圆屋顶
9. 带走
10. 倾向，趋势

Down（竖排）

1. 延展
2. 系数
3. 黏度
6. 下降
7. 捕汁器
8. 冷凝，浓缩

Task 2: One-choice question

根据对话内容，选择下列问题的正确答案。

1. The evaporation process in the sugar industry is conducted using ________.
   A. water B. sugar
   C. steam D. juice
2. The steam-heating unit in the multiple effect evaporator is called ________.
   A. calandria B. vacuum pump
   C. effect D. tube
3. The boiling juice rises in the tubes and releases ________.
   A. vapors B. crystals
   C. impurities D. non-sugars
4. The vacuum in the evaporator causes the liquor to flow from ________.
   A. top to bottom
   B. bottom to top
   C. effect to effect
   D. one effect to the following one
5. Impurities in the juice are removed during ________.
   A. crystallization B. evaporation
   C. supersaturation D. filtration
6. Entrainment refers to the condition where ________.
   A. steam is introduced into the calandria
   B. juice rapidly decomposes
   C. the liquor level is incorrect
   D. small droplets of liquor are carried by vapors
7. The vacuum is maintained by condensing the vapors using ________.
   A. cooling water B. steam
   C. sugar D. pump
8. The evaporation process continues until ________.
   A. the juice decomposes
   B. all the water is evaporated
   C. crystals form and separate
   D. the viscosity of the liquor increases

Task 3: Blank filling

根据对话内容，把下列句子补充完整，每空填写一个单词。

1. Evaporation is conducted by means of ________ in evaporators.
2. A multiple effect evaporator consists of a series of large cylindrical ________.
3. The ________ is a drum-shaped steam chest provided with a large vertical central tube.
4. The ________ of the first effect receives the thin juice and exhaust or low pressure steam.
5. The boiling juice in the tubes rises, releases vapors, and leaves the effect through the ________.
6. The liquor then descends through the "down take" in which no ________ takes place.
7. The vacuum is maintained in each effect by ________ the vapors.
8. The degree of supersaturation is conveniently expressed by the coefficient of ________.

Task 4: Thinking training

根据文章内容，把下列关于多效蒸发器的表格填写完整，并尝试与搭档一起解释多效蒸发器的工作过程。

Features	Description
Composition of the multiple effect evaporator 多效蒸发器的构成	1. A multiple effect evaporator in the sugar industry consists of a series of large ________. 2. All of the effects in a multiple effect evaporator are of similar construction and design, consisting of a large ________ or metal plate cylindrical vessel.
Working process of multiple effect evaporator 多效蒸发器工作过程	3. ________ and ________ are introduced into the calandria of the first effect. 4. Effective heat is transferred to the ________, causing it to boil. 5. The boiling juice rises in the tubes, releases the ________ which leave the effect through the dome. 6. The liquor then ________ at the center of the effect through the "down take". 7. The ________ maintained in each of the effects is the force that causes the liquor to flow. 8. The rate of flow being controlled by ________. 9. In the case of the last effect, the concentrated liquor is ________ by means of a pump, and is ready for final evaporation.

Key (Unit 5 Section 2)

Task 1:

	1 e														
	x														
	t														
	e			2 c							3 v				
	n			o		4 s	o	l	u	b	i	l	i	t	y
	5 d	o	m	e							s				
				f							c				
				f							o				
				i		6 d					s				
7 c		8 c		c		9 e	n	t	r	a	i	n			
a		o		i		s					t				
10 t	e	n	d	e	n	c	y				y				
c		d		n		e									
h		e		t		n									
-		n				d									
a		s													
l		e													
l															

Task 2:

1. C 2. A 3. A 4. D 5. D 6. D 7. A 8. C

Task 3:

1. steam 2. vessels 3. calandria 4. calandria 5. dome 6. heating 7. condensing 8. supersaturation

Task 4:

1. cylindrical vessels 2. vertical cast-iron 3. Exhaust, low pressure steam 4. juice 5. vapors 6. descends 7. vacuum 8. valve 9. removed

译文

第二节：蒸发过程

澄清站产生的稀汁液需立即进行蒸发，生产结晶糖。这一操作不能延误，因为稀汁液会迅速分解，造成糖分损失。蒸发操作是通过蒸汽在蒸发罐中实现的，在第一阶段，汁液被浓缩到 50% ~ 60%。这个阶段使用多效蒸发器进行，其能够非常高效地利用蒸汽。

多效蒸发器：

在糖厂中，多效蒸发器由一系列大型圆柱形容器组成（通常使用三个或四个蒸发罐，分别称为三效或四效）。多效蒸发器的所有效应都采用相似的结构和设计，由一个大型垂直的铸铁或金属板圆柱形容器组成。在糖业中，蒸汽加热部件（通常是一个冷却管）放置在底部。冷却管是一个圆筒形的蒸汽箱，内有一个大型垂直中央管道（下管），从顶部延伸到底部滚筒叶片。在滚筒的其他部分，有许多小直径的垂直管道，也从顶部延伸到底部叶片。

废气或低压蒸汽被导入第一反应筒中，该反应筒也接收稀汁。蒸汽中的有效热量通过加热管道壁传递给糖汁，使其沸腾。沸腾的汁液上升至管道中，释放蒸汽通过圆屋顶与管道输送给下一个反应冷却器。然后，液体在反应中心通过“下管”下降，此处不加热。建立的自然循环的有效性取决于正确设计的反应和液体的黏度。随着蒸发过程的进行，黏度增加，因此在最后一个效应中循环略微较差。

每个蒸发罐中保持真空是导致液体从一个反应罐流向下一罐反应的原因，流速由阀门控制。在末效罐中，浓缩后的糖浆通过泵移送，准备在真空罐中进行最后的蒸发。

在水分去除的浓缩过程中，即使是最清澈的糖浆也会包含不溶解物质。一些沉淀物仍然存在于糖浆中，使其变得混浊，必须通过后续的过滤来去除。其他杂质会在管道的加热表面上形成积垢，特别是在末效罐中，这种结垢必须定期清除，因为它会减慢热量传递并影响蒸发器的效率。

离开汁液表面的蒸汽具有一定的速率，可能携带小水滴，这种情况被称为“雾沫夹带”。通过适当设计反应并保持汁液的量正确可以避免带液现象。每个反应的圆顶中都内置了一个称为“捕汁器”的挡板，用于从蒸汽中分离水滴。

末效罐通过冷却水冷凝产生的蒸汽来保持真空。在其他蒸发罐中，通过连续反应的排管中冷凝蒸汽而保持真空。真空的适宜度由泵、隔栏或气压排水来控制，这些设备用于去除冷凝器中的冷凝水，并通过真空泵去除不可凝结气体。

结晶：

在这个阶段，继续蒸发直到糖晶体形成并从剩余的水和杂质中分离出来。

超饱和溶液的作用：

在不同温度下，蔗糖在水中溶解度的数据见表 5-1。

表 5-1　蔗糖在水中的溶解度

温度/℃	30	40	50	60	70	80
质量比/%	68.70	70.42	72.25	74.18	16.22	78.36

如果冷却饱和蔗糖溶液，或者蒸发一些水，溶液就会变得过饱和，立即结晶。过饱和程度通过饱和系数方便地表示，该系数是通过将溶液中的糖分质量除以该温度下的理论溶解度获得的。蔗糖在过饱和溶液中通常具有结晶的天然趋势，但这需要一定的时间，时间长短取决于各种条件。当过饱和系数较小时，即只存在轻微过饱和时，蔗糖只会结晶在已经存在的溶液中。如果没有这样的晶体存在，溶液可能会持续超饱和很长时间。然而，当溶液强烈过饱和时，例如，过饱和系数为 1.5，蔗糖会自发结晶。还存在一种中间过饱和区域，当存在已有晶体时，新的晶体（伪晶）会形成，在没有晶体的情况下则不会形成。

过饱和度对于引发结晶起到了正向作用，但也有一些负面影响会抑制结晶，其中之一是黏度。为了实现自发结晶，必须存在更高的过饱和系数。糖厂糖浆不仅包含纯蔗糖溶液，还包含非糖物质，这些物质不仅增加了黏度，还改变了蔗糖的溶解度。糖浆中的某些非糖物质明显增加了蔗糖的溶解度，而其他非糖物质则具有降低蔗糖溶解度的相反特性。糖浆中的非糖物质在煮沸过程中具有非常明确的技术重要性。

Section 3 Identify the picture

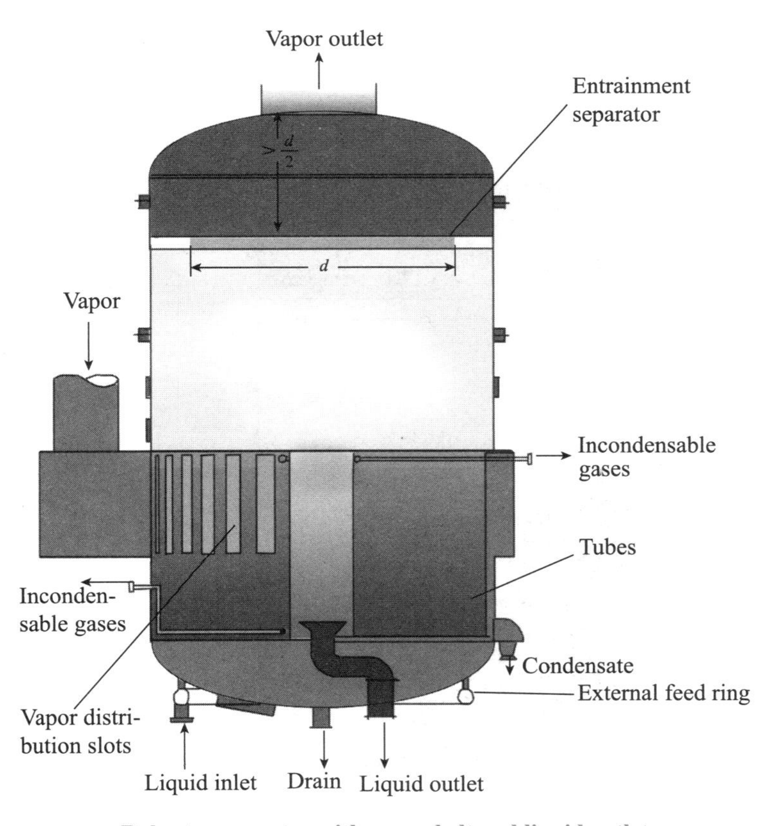

Robert evaporator with vapor belt and liquid outlet

New words and expressions

1. vapor	/ˈveɪpər/	*n.* 蒸汽
2. drain	/dreɪn/	*n.* 排出口
3. condensate	/ˈkɑːndənˌseɪt/	*n.* 冷凝水
4. tubes	/tjuːbz/	*n.* 加热管
5. vapor outlet		蒸汽出口
6. incondensable gases		不凝气体
7. vapor distribution slot		蒸汽分配口
8. liquid inlet		糖汁入口
9. liquid outlet		糖汁出口
10. external feed ring		外部进料环
11. incondensable gases		不凝气体
12. entrainment separator		雾沫分离器

Exercises

Task1: Fill in the blanks

认真观察图片，并在空白处填上正确的单词

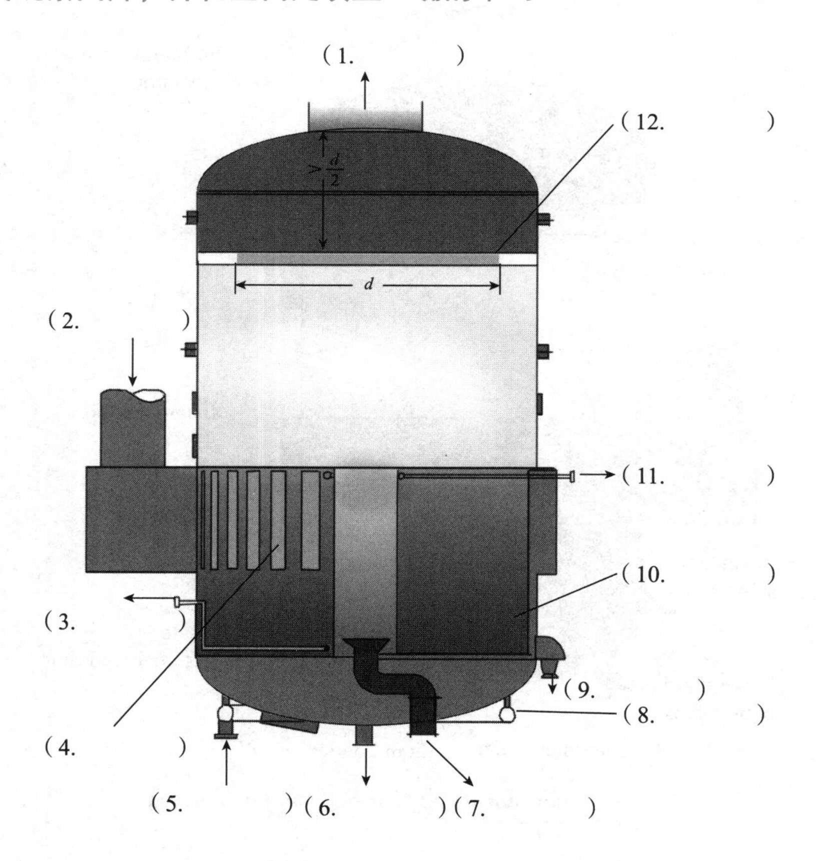

Key（Unit5　Section 3）

Task1:

1. vapor outlet 蒸汽出口	2. vapor 蒸汽
3. incondensable gases 不凝气体	4. vapor distribution slot 蒸汽分配口
5. liquid inlet 糖汁入口	6. drain 排出口
7. liquid outlet 糖汁出口	8. external feed ring 外部进料环
9. condensate 冷凝水	10. tubes 加热管
11. incondensable gases 不凝气体	12. entrainment separator 雾沫分离器

Culture Background

Furnace of "Pin" Character Shape

China began using porous brick furnaces as early as the Han Dynasty, and the *Tian Gong Kai Wu* records the use of the "Pin" character furnace for evaporating sugar juice in China. This technology is an important innovation in the world's sugar industry, just like a vertical roller mill, which quickly spread throughout the world.

"品"字火炉

中国早在汉朝就开始使用多孔砖炉，《天工开物》则记载了中国使用"品"字火炉用于蒸发糖汁。这项技术是世界制糖业的重要革新，就像垂直的碾辊压榨机，迅速传遍全世界。

Unit 6 Crystallization

Objectives

After learning this unit, you will be able to

1. Expand your vocabulary about crystallization;
2. Understand the purpose and the process of crystallization;
3. Summarize the basic operation of the crystallization process.

Section 1 Discussing the purpose and basic operation of the crystallization process

Characters:

Mr. Sweet—a sugar industry professional （制糖行业的专业人士）

Zhang Jiejing, Tang Qianxi and Lin Huahua—students majoring in Sugar Engineering and being enthusiastic about crystallization process（糖业工程专业学生，对甘蔗糖厂中的结晶过程有浓厚兴趣）

Mr. Sweet: Hello class, today we're going to talk about the crystallization process in a sugarcane sugar factory. Can anyone tell me what the purpose of crystallization is?

Zhang Jiejing: Is it to separate the sugar from the **remaining**[1] water and impurities?

Mr. Sweet: Exactly! The crystallization process is carried out to separate the sugar crystals from the syrup that is formed during the evaporation process. Now, does anyone know how this process works?

Tang Qianxi: I think it involves creating a supersaturated solution to encourage the sugar to crystallize.

Mr. Sweet: Great job! During the evaporation process, the sugar solution becomes supersaturated, meaning it has more sugar dissolved in it than it would under normal conditions. This supersaturation provides the conditions necessary for sugar crystals to form spontaneously. Now, let's discuss the basic operation of the crystallization process.

Lin Huahua: So, how is the sugar syrup transformed into crystals?

Mr. Sweet: Good question! The first step is to transfer the syrup from the evaporation stage into large tanks known as crystallizers. Inside these tanks, the syrup is cooled and stirred to encourage the formation of sugar crystals. This cooling process helps to further increase the supersaturation level of the syrup.

Zhang Jiejing: What happens after the sugar crystals start to form?

Mr. Sweet: Once the sugar crystals begin to form, they need to be separated from the remaining liquid. This is done through a process called separation. The mixture of sugar crystals and liquid is sent to a **centrifuge**[2], which spins rapidly to separate the crystals from the syrup. The separated crystals are then dried to remove any remaining moisture, resulting in the production of crystal sugar.

Tang Qianxi: That's interesting! So, the purpose of crystallization is not only to remove impurities but also to transform the syrup into usable sugar crystals.

Mr. Sweet: Absolutely! The crystallization process not only purifies the sugar but also allows it to be transformed into a solid form that can be easily stored and used. And that's the basic concept behind the crystallization process in a sugarcane sugar factory.

Lin Huahua: Thank you, teacher! This was really informative.

Mr. Sweet: You're welcome! I'm glad you found it helpful. Do you have any more questions on this topic?

Zhang Jiejing: How long does the crystallization process usually take?

Mr. Sweet: The duration of the crystallization process can vary depending on several factors, such as the temperature, sugar concentration, and stirring intensity. It can take several hours to a few days to complete.

Tang Qianxi: Alright, got it! Thanks for clarifying.

Mr. Sweet: You're welcome! If you have any more questions in the future, feel free to ask.

Lin Huahua: Sure!

音频：Unit 6 Section 1

New words and expressions

1. remaining /rɪ'meɪnɪŋ/ *adj.* 剩余的
2. centrifuge /'sentrɪfjuːdʒ/ *n.* 离心机

Exercises

Task 1: Crosswords puzzle

请根据中文提示把下面字谜中的单词填写出来。

Across（横排）

3. 分开，分离
4. 杂质
6. 使过度饱和
8. 蒸发
10. 浓度

Down（竖排）

1. 离心机
2. 糖浆；糖水
5. 结晶
7. 剩余的
9. 潮气

Task 2: One-choice question

根据对话内容，选择下列问题的正确答案。

1. What is the purpose of the crystallization process?________
   A. To separate water from sugar crystals.
   B. To transform syrup into usable sugar crystals.
   C. To remove impurities from the sugar solution.
   D. All of the above.
2. What is the passive form of the sentence "They produce crystal sugar in the factory"?________
   A. Crystal sugar is being produced in the factory.
   B. Crystal sugar will be produced in the factory.
   C. Crystal sugar has been produced in the factory.
   D. Crystal sugar was produced in the factory.
3. What does the term "supersaturated" mean?________
   A. A solution that has less solute than could be dissolved.
   B. A solution that has an equal amount of solute and solvent.
   C. A solution that contains more solute than can normally be dissolved.
   D. A solution that contains no solute at all.

Task 3: Blank filling

根据对话内容，把下列句子补充完整，每空填写一个单词。

1. During the crystallization process, the sugar crystals are separated from the ________.
2. The ________ of the sugar solution is increased through the evaporation process.
3. The syrup is cooled and stirred in large tanks called ________.
4. Crystallization helps to ________ the sugar and remove impurities.
5. The ________ of the sugar syrup is crucial for successful crystallization.

Task 4: Thinking training

根据对话内容，把结晶的目的和结晶过程用思维导图的形式绘制出来，并复述。

Purpose ⟷ Crystallization ⟷ Process

The crystallization process is carried out to ________________________________

The first step of crystallization is to ________________________________

____

Key (Unit 6 Section 1)

Task 1:

<table>
<tr><td></td><td></td><td></td><td></td><td></td><td>1 c</td><td></td><td></td><td></td><td></td><td></td><td></td><td></td><td></td><td></td><td></td><td></td><td></td><td></td><td></td><td></td></tr>
<tr><td></td><td></td><td>2 s</td><td></td><td>3 s</td><td>e</td><td>p</td><td>a</td><td>r</td><td>a</td><td>t</td><td>i</td><td>o</td><td>n</td><td></td><td></td><td></td><td></td><td></td><td></td><td></td></tr>
<tr><td></td><td></td><td>y</td><td></td><td></td><td>n</td><td></td><td></td><td></td><td></td><td></td><td></td><td></td><td></td><td></td><td></td><td></td><td></td><td></td><td></td><td></td></tr>
<tr><td></td><td></td><td>r</td><td></td><td></td><td>t</td><td></td><td></td><td></td><td></td><td></td><td></td><td></td><td></td><td></td><td></td><td></td><td></td><td></td><td></td><td></td></tr>
<tr><td></td><td></td><td>u</td><td></td><td></td><td>r</td><td></td><td></td><td></td><td></td><td></td><td></td><td></td><td></td><td></td><td></td><td></td><td></td><td></td><td></td><td></td></tr>
<tr><td>4 i</td><td>m</td><td>p</td><td>u</td><td>r</td><td>i</td><td>t</td><td>y</td><td></td><td></td><td></td><td></td><td></td><td></td><td></td><td></td><td></td><td></td><td></td><td></td><td></td></tr>
<tr><td></td><td></td><td></td><td></td><td></td><td>f</td><td></td><td></td><td></td><td></td><td></td><td></td><td></td><td>5 c</td><td></td><td></td><td></td><td></td><td></td><td></td><td></td></tr>
<tr><td></td><td></td><td></td><td></td><td>6 s</td><td>u</td><td>p</td><td>e</td><td>7 r</td><td>s</td><td>a</td><td>t</td><td>u</td><td>r</td><td>a</td><td>t</td><td>e</td><td></td><td></td><td></td><td></td></tr>
<tr><td></td><td></td><td></td><td></td><td></td><td>g</td><td></td><td></td><td>e</td><td></td><td></td><td></td><td></td><td>y</td><td></td><td></td><td></td><td></td><td></td><td></td><td></td></tr>
<tr><td></td><td></td><td></td><td></td><td></td><td>e</td><td></td><td></td><td>m</td><td></td><td></td><td></td><td></td><td>s</td><td></td><td></td><td></td><td></td><td></td><td></td><td></td></tr>
<tr><td></td><td></td><td></td><td></td><td></td><td></td><td>8 e</td><td>v</td><td>a</td><td>p</td><td>o</td><td>r</td><td>a</td><td>t</td><td>i</td><td>o</td><td>n</td><td></td><td></td><td></td><td></td></tr>
<tr><td></td><td></td><td></td><td></td><td></td><td></td><td></td><td></td><td>i</td><td></td><td></td><td></td><td></td><td>a</td><td></td><td></td><td></td><td></td><td></td><td></td><td></td></tr>
<tr><td></td><td></td><td></td><td></td><td></td><td></td><td></td><td></td><td>n</td><td></td><td></td><td></td><td></td><td>l</td><td></td><td></td><td></td><td></td><td></td><td></td><td></td></tr>
<tr><td></td><td></td><td></td><td></td><td></td><td></td><td></td><td></td><td>i</td><td></td><td></td><td></td><td></td><td>l</td><td></td><td></td><td></td><td></td><td></td><td></td><td></td></tr>
<tr><td></td><td></td><td></td><td></td><td></td><td></td><td></td><td></td><td>n</td><td></td><td></td><td></td><td></td><td>i</td><td></td><td>9 m</td><td></td><td></td><td></td><td></td><td></td></tr>
<tr><td></td><td></td><td></td><td></td><td></td><td></td><td></td><td></td><td>g</td><td></td><td></td><td></td><td></td><td>z</td><td></td><td>o</td><td></td><td></td><td></td><td></td><td></td></tr>
<tr><td></td><td></td><td></td><td></td><td></td><td></td><td></td><td></td><td></td><td></td><td></td><td></td><td></td><td>a</td><td></td><td>i</td><td></td><td></td><td></td><td></td><td></td></tr>
<tr><td></td><td></td><td></td><td></td><td></td><td></td><td></td><td></td><td></td><td></td><td></td><td></td><td></td><td>t</td><td></td><td>s</td><td></td><td></td><td></td><td></td><td></td></tr>
<tr><td></td><td></td><td></td><td></td><td></td><td></td><td></td><td></td><td></td><td></td><td></td><td></td><td></td><td>i</td><td></td><td>t</td><td></td><td></td><td></td><td></td><td></td></tr>
<tr><td></td><td></td><td></td><td></td><td></td><td></td><td></td><td></td><td></td><td></td><td></td><td></td><td></td><td>o</td><td></td><td>u</td><td></td><td></td><td></td><td></td><td></td></tr>
<tr><td></td><td></td><td></td><td></td><td></td><td></td><td></td><td></td><td>10 c</td><td>o</td><td>n</td><td>c</td><td>e</td><td>n</td><td>t</td><td>r</td><td>a</td><td>t</td><td>i</td><td>o</td><td>n</td></tr>
<tr><td></td><td></td><td></td><td></td><td></td><td></td><td></td><td></td><td></td><td></td><td></td><td></td><td></td><td></td><td></td><td>e</td><td></td><td></td><td></td><td></td><td></td></tr>
</table>

Task 2:

1. D 2. A 3. C

Task 3:

1. syrup 2. concentration 3. crystallizers 4. purify 5. supersaturation

Task 4:

The purpose of crystallization is to separate sugar crystals from the syrup formed during the evaporation process.

The crystallization process in a sugarcane sugar factory is as follows:

First, during the evaporation process, the sugar solution becomes supersaturated. Then, the syrup is transferred from the evaporation stage to large tanks called crystallizers. Inside the crystallizers, the syrup is cooled and stirred to encourage the formation of sugar crystals. Once the sugar crystals start to form, the mixture of sugar crystals and liquid is sent to a centrifuge to separate the crystals from the syrup. Finally, the separated crystals are dried to remove any remaining moisture, resulting in the production of crystal sugar. The duration of the crystallization process can vary depending on factors like temperature, sugar concentration, and stirring intensity.

译文

第一节：讨论结晶目的和基本操作过程

甜先生：大家好，今天我们要谈谈甘蔗糖厂中的结晶过程。有人能告诉我结晶的目的是什么吗?

张结晶：是不是为了将糖从剩余的水和杂质中分离出来?

甜先生：没错！结晶过程就是将糖晶体从蒸发过程中形成的糖浆中分离出来。那么，有人知道这个过程是如何进行的吗?

唐千禧：我认为它涉及创建一个过饱和溶液，以促使糖结晶。

甜先生：非常好！在蒸发过程中，糖溶液变得过饱和，意味着其中溶解的糖比正常条件下要多。这种过饱和提供了糖结晶自发形成所需的条件。那么，接下来我们来讨论结晶过程的基本操作。

林华华：那么，糖浆是如何转变为结晶的呢?

甜先生：好问题！首先，将糖浆从蒸发阶段转移到大型的结晶槽中。在这些槽中，通过冷却和搅拌来促使糖晶体的形成。这个冷却过程有助于进一步提高糖浆的过饱和度。

张结晶：糖晶体开始形成后会发生什么?

甜先生：一旦糖晶体开始形成，就需要将其与剩余的液体分离开来。这是通过一个称为分离的过程来完成的。将糖晶体和液体混合物送入离心机，离心机高速旋转以将晶体与糖浆分离。分离出的糖晶体随后被干燥以去除任何剩余的水分，从而生产出结晶糖。

唐千禧：很有趣！所以，结晶的目的不仅是去除杂质，还是将糖浆转变为可用的糖晶体。

甜先生：完全正确！结晶过程不仅可以提纯糖，还可以将其转化为易于储存和使用的固态形式。这就是甘蔗糖厂中结晶过程的基本概念。

林华华：谢谢老师！这真的很有帮助。

甜先生：不客气！我很高兴你觉得有用。对这个话题还有其他问题吗?

张结晶：结晶过程通常需要多长时间?

甜先生：结晶过程的时间长度可以因多种因素而异，如温度、糖浓度和搅拌强度。完成这个过程可能需要几个小时到几天的时间。

唐千禧：好的，明白了！谢谢您的解释。

甜先生：不客气！如果将来还有其他问题，请随时提问。

林华华：会的。

Section 2 The boiling operation

The boiling operation which may be heated either by coils（蛇管；线圈）or by a calandria unit（汽鼓装置）. **In appearance**[1] and **in** a certain other **respects**[2] vacuum pan are quiet similar to single effect evaporators（单效蒸发器）.

There are many methods, and also, modifications of each method for conducting the boiling operation. This fact must be **kept in mind**[3] in reading the follow discussion.

Different Methods of Operation:

In one method, one operator first adds a portion of evaporator syrup to the pan. The amount to be added is based on what the operator considers will be sufficient to produce, the correct number of crystal nuclei or as it is term "to form the proper grain". The graining charge varies in extent, but is frequently equal to from one-fourth to one-third the full capacity of the pan. After this charge is added to the pan, a vacuum is established, and steam is admitted to the heating unit, that is, to the coil and calandria. The liquor is **thereby**[4] concentrated until it reaches the degree supersaturation where crystal will form. There are several ways of conducting this stage of the operation. Some operator use the so-called "waiting method" in which the liquor in the pans is kept in a high degree of supersaturation until the number of crystals are form. Other operators employ what is known as a "**shock**"[5] which can be produced **in various ways**[6]. One method is to boil the liquor at not so high vacuum, in **consequently**[7], at a high temperature. When the liquor reached the desired degree of supersaturation the vacuum is rapidly in increased, and the follow of the steam through the heating units is reduced. The temperature in the pan is thereby **lowered**[8], and results in greater supersaturation due to the reduced solubility of sugar at low temperature. In other cases, the shock is **administered**[9] by drawing in a small amount of cold juice. Often if air is added to the pan, the dust present will induce the formation of nuclei. A shock method that many prefer is to draw some **powdered**[10] sugar into the pan immediately after the metastable stage has been reached, and wait for the development which comes gradually and **smoothly**[11]. The shocking must to be done too late as this results in forming, not only nuclei, but **conglomerates**[12] as well.

The graining must be carefully controlled, since, when too few crystals are form, there is not sufficient crystal surface for the sugar to deposit upon, it will result in a **infinite**[13] variety of crystal sizes. This small grains **plug**[14] up the **drainage**[15] between the large ones, and thus **make for**[16] difficult centrifugal work.

Seeding[17]:

A procedure known as seeding provides a very effective method of conducting the graining operation. In this the full and proper number of minute sugar crystals is introduced into the pan as soon as saturation has been **exceeded**[18], and the concentration is not allowed to reach the point where new crystals will for.

Building the Crystals:

After the grain is established in the pan by any of the various methods, the supersaturation is **henceforth**[19] kept below the point where additional crystals known as "false grain" will form. This is accomplished by admitting more liquor to the pan, either **intermittently**[20] or continuously, although a continuous feed is **preferable**[21]. **From here on**[22] a balance must be struck between the rate of evaporation and the amount of liquor that is admitted to the pan. Evaporation must proceed at a rate that will maintain a supersaturation sufficient to cause rapid growth of the crystals. On the other hand, the evaporation must not exceed the rate at which sugar can deposit on existing crystals since this would lead to a much higher supersaturation at which "false grain" will form.

The whole mass must be kept in motion since each crystal rapidly depletes the sucrose from the surrounding liquor, and must be continually brought into contact with a fresh liquor phase. Circulation is also necessary to prevent **local**[23] over-heating of the portion of the liquor in contact with the **heating coils**[24]. **Apart from**[25] any decomposition effect of sugar, local overheating would also produce a case of local **undersaturation**[26] of the sugar.

At first the natural circulation is reasonably good, but **after a time**[27] the **fluidity**[28] of the mass decreases. This is due to the fact that the growing crystals soon occupy **considerable**[29] volume, and the fluid space becomes relatively small. Various methods exist for promoting circulation. The **incoming**[30] liquor flows upward and promotes circulation.

Steam may be admitted to the bottom of the pan, and as the steam bubbles rise under vacuum conditions they expand, and in that way assist the circulation. Much study has been given to the influence of the shape of the pan, and as a result of these studies very definite improvement has been made in pan design.

Where, who has done much to clarify our under-standing of sugar boiling, has designed a slowly revolving screw placed in the central tube which **boosts**[31] the circulation, in exactly the same direction as the original natural circulation, namely, up the outer **periphery**[32] and down the center. This device has been **incorporated**[33] into some types of vacuum pans. While many people believe the benefits of mechanical circulation are more apparent when obtaining the second and third crop of crystals, actually, very advantageous results have been accomplished from the use of mechanical circulators（机械循环器）on refinery liquors of high purity. In this latter case, although the gain of speed is not very great, the formation and the color of the crystals is very much better.

Discharging[34] the Pan:

When the crystallization has proceeded to the point where, **for practical purposes**[35], it is considered complete, the contents of the pan are discharged, or as it is termed a "strike" is made. The mixture of crystals and mother liquor is termed a "massecuite" or a "fill mass". This massecuite or fill mass then goes to the centrifugal machines where the mother liquor is spun off from the crystals.

Multiple Boiling:

The mother liquor that is **spun**[36] off is called a "A molasses" in cane sugar factories, or a "**high green syrup**"[37] in beet sugar factories. This liquor still contains much **crystallizable**[38] sugar which can be recovered by further boiling. Some factories employ a two-boiling system, but the three boiling system is more common. In a three-boiling system, the liquor from the first or "A" crystallization is reboiled in the second vacuum pan, known as the intermediate pan in which a second or "B" crop of crystals is produced. The mother liquor from the B crop of crystals is again boiled to produce a third or "C" crop of crystals. The mother liquor from the C crystals constitutes the final molasses.

Obtaining the "C" Crystals:

During the course of this process the non-sugars become more concentrated in the liquor, and increase the difficulties in crystallization. This difficulty becomes particularly **pronounced**[39] in the third boiling, and since, in many factories, this is the last chance to recover the sugar, much study has been given to this stage.

The procedure generally employed for obtaining the third or "C" crop of crystals involves two steps. The manner of conducting the first step varies in different plants, but one satisfactory way is to boil a relatively small quantity of high purity syrup to form the grain. After the grain has formed the mother liquor from the second, or "B" crystals is then admitted to the pan, and the boiling is continued until it is strongly supersaturated. The supersaturated massecuite is then discharged to a crystallizer in which the crystallization proceeds to completion.

音频：Unit 6　Section 2

New words and expressions

1. in appearance		看上去，表面上，外观上
2. in… respect		在……方面
3. keep in mind		记住，考虑到
4. thereby	/ˌðeəˈbaɪ/	*adv.* 因此，从而
5. shock	/ʃɒk/	*n.* 震动，刺激
6. in various ways		用多种方法

No.	Word/Phrase	Pronunciation	Meaning
7.	consequently	/ˈkɒnsɪkwəntli/	*adv.* 因此
8.	lower	/ˈləʊə(r)/	*v.* 减少，降低
9.	administer	/ədˈmɪnɪstə(r)/	*v.* 进行，实施
10.	powdered	/ˈpaʊdəd/	*adj.* 变成粉末的
11.	smoothly	/ˈsmuːðli/	*adv.* 平稳地，连续而流畅地
12.	conglomerate	/kənˈglɒmərət/	*v.* 聚结，聚集；堆积成团
13.	infinite	/ˈɪnfɪnət/	*adj.* 无限的
14.	plug	/plʌg/	*v.* 堵住，塞住，堵塞
15.	drainage	/ˈdreɪnɪdʒ/	*n.* 排水系统；排水
16.	make for		有利于
17.	seeding	/ˈsiːdɪŋ/	*n.* 晶种
18.	exceed	/ɪkˈsiːd/	*v.* 超过，超出
19.	henceforth	/ˌhensˈfɔːθ/	*adv.* 从此之后
20.	intermittently	/ˌɪntəˈmɪtəntli/	*adv.* 间歇地
21.	preferable	/preferable/ˈprefrəb(ə)l/	*adj.* 更好的
22.	from here on		此后，从这里开始
23.	local	/ˈləʊk(ə)l/	*adj.* 局部的，当地的
24.	heating coil		加热蛇管（加热线圈）
25.	apart from…		除……以外
26.	undersaturation	/ˈʌndəsætʃəˈreɪʃən/	*n.* 未饱和
27.	after a time		一些时候以后
28.	fluidity	/fluˈɪdəti/	*n.* 流动性
29.	considerable	/kənˈsɪdərəb(ə)l/	*adj.* 相当大的；大量的
30.	incoming	/ˈɪnkʌmɪŋ/	*n.* 进料，加入
31.	boost	/buːst/	*v.* 使增长，推动 *n.* 推动，促进
32.	periphery	/pəˈrɪfəri/	*n.* 外围，边缘
33.	incorporate	/ɪnˈkɔːpəreɪt/	*v.* 合并；混合
34.	discharging	/disˈtʃaːdʒiŋ/	*v.* 卸货；排出 *n.* 卸货
35.	for practical purposes		从实际上看
36.	spin	/spɪn/	*v.* （使）快速旋转
37.	high green syrup		甲原蜜
38.	crystallizable	/krɪstəˈlaɪzəbl/	*adj.* 可结晶的
39.	pronounced	/prəˈnaʊnst/	*adj.* 显著的，明显的

Exercises

Task 1: Crosswords puzzle

请根据中文提示把下面字谜中的单词填写出来。

Across（横排）

1. 震动，刺激
4. 流动性；流质
6. 管理，治理
7. 因此，结果
9. 晶种
10. 未饱和

Down（竖排）

2. 可结晶的
3. 变成粉末的
5. 聚结，聚集
8. 排水系统，排水

Task 2: One-choice question

根据对话内容，选择下列问题的正确答案。

1. The boiling point of water is ________ degrees Celsius.

A. 0　　B. 100　　C. 200　　D. 300

2. The ________ of the liquid caused it to evaporate quickly.

A. boiling B. freezing C. melting D. condensing

3. We need to adjust the ________ of the solution to make it more concentrated.

A. pressure B. temperature C. concentration D. vaporization

4. The crystals formed through the process of ________.

A. freezing B. evaporation C. precipitation D. crystallization

5. The solubility of sugar in water is very ________.

A. low B. high C. stable D. unknown

6. The syrup is ________ with a concentrated flavor.

A. sweet B. salty C. bitter D. sour

7. The ________ of the solution increased as more sugar was added.

A. freezing point B. boiling point C. saturation D. evaporation

8. The ________ of the liquid determines how much solute it can dissolve.

A. concentration B. precipitation C. nucleation D. viscosity

Task 3: Blank filling

根据对话内容，把下列句子补充完整，每空填写一个单词。

1. The process of turning a liquid into a gas is called ________.
2. The ________ of the solution increased as more solute was added.
3. The crystals in the solution were formed through a process known as ________.
4. The ________ of the solution was too high, leading to the formation of false grain.
5. The ________ method is a common way of conducting the graining operation.
6. The ________ of sugar can be affected by temperature and pressure.
7. The ________ of the liquid in the pan is maintained through evaporation.
8. The ________ of sugar can be recovered by further boiling.

Task 4: Reading comprehension

根据文章内容，回答下列问题。

1. According to the passage, which ways are mentioned to concentrate the liquor and make it reach the degree supersaturation where crystal will form?

____

____

2. To clarify our under-standing of sugar boiling, what has been designed and placed in the central tube to boost the circulation?

____

____

Key (Unit 6 Section 2)

Task 1:

							1 s	h	o	2 c	k										
										r								3 p			
			4 f	l	u	i	d	i	t	y								o			
										s				5 c				w			
										t				o				d			
										6 a	d	m	i	n	i	s	t	e	r		
										l				g				r			
7 c	o	n	s	e	q	u	e	n	t	l	y			l				e			
										i				o				d			
								8 d		z				m							
								r		a		9 s	e	e	d	i	n	g			
								a		b				r							
								i		l				a							
							10 u	n	d	e	r	s	a	t	u	r	a	t	i	o	n
								a						e							
								g													
								e													

Task 2:

1. B 2. A 3. C 4. D 5. B 6. A 7. C 8. A

Task 3:

1. evaporation 2. concentration 3. crystallization 4. supersaturation 5. seeding 6. solubility 7. supersaturation 8. molasses

Task 4:

1. Waiting method and shock method. 2. A slowly revolving screw .

译文

第二节：煮糖工艺

煮糖工艺可以通过蛇管或汽鼓装置加热。外观上和某些其他方面，煮糖罐与单效蒸发器相似。

有许多方法以及每种方法的修改用于进行煮糖工艺。在阅读下面的论述时必须记住这一点。

不同的操作方法：

其中一种方法，煮糖工人首先将少量的蒸发糖浆加入煮糖罐中。添加的量基于工人生产效率，能够产生正确数量的晶核或所谓的“形成适当的颗粒”的需求。形成颗粒材料的量不同，但通常相当于煮糖罐容量的四分之一到三分之一。在将原料加入煮糖罐中后，抽真空，并通入蒸汽到加热单元，即线圈和汽鼓。从而溶液被浓缩，直到达到结晶。有几种方法来进行这个操作阶段。一些操作员使用所谓的“等待法”，在这种方法中，煮糖罐中的溶液保持高度过饱和状态，直到结晶数量满足。其他操作员使用所谓的“震动”方法，“震动”可以用多种方法产生。一种方法是在不那么高的真空条件下沸腾溶液，因此所需温度较高。当溶液达到所需的过饱和度时，真空迅速增加，并减少通过加热单元的蒸汽流量。煮糖罐中的温度因此降低，由于低温下蔗糖的溶解度降低而产生更高的过饱和度。在其他情况下，通过向煮糖罐中加入少量的冷汁来进行刺激。通常，如果向煮糖罐中注入空气，其中存在的灰尘将引发结晶核的形成。许多人偏爱的刺激方法是在达到亚稳态阶段后立即向煮糖罐中加入一些粉状糖糊，并等待逐渐平稳地发展。刺激不能太晚进行，因为这将导致形成的不仅是结晶核还有团块。

必须仔细控制结晶，因为如果形成的结晶太少，糖无法在足够的结晶面上沉积，将会导致各种大小的结晶。这些小颗粒会阻塞大颗粒之间的排液通道，从而使离心机工作更加困难。

晶种：

晶种的程序为进行形成颗粒操作提供了一种非常有效的方法。一旦超饱和度被超过，立即向煮糖罐中引入全部和适量的微小糖晶体，不允许浓度达到产生新晶体的程度。

养晶过程：

在煮糖罐中通过各种方法生成晶体后，过饱和度将保持在不形成“伪晶”的点以下。这是通过向煮糖罐中间歇或连续地加入更多的溶液来实现的，尽管连续进料更可取。从此时开始，必须在蒸发速率和加入煮糖罐中的溶液量之间取得平衡。蒸发速率必须保持在足以使晶体迅速生长的超饱和度水平，但又不能超过糖在现有晶体上的沉积速率，否则这将导致更高的超饱和度，产生“伪晶”。

整个物料必须保持运转，因为每个晶体很快耗尽周围溶液中的蔗糖，必须不断与新鲜液相接触。循环也是必要的，以防止与加热线圈接触的部分液相过热。除了糖的任何分解效应外，局部过热也会导致糖的局部未饱和。

起初，自然循环相对较好，但一段时间后物料的流动性降低。这是因为持续生长的晶体很快占据了相当大的体积，流动空间相对较小。存在各种方法来促进循环。加入的溶液向上流动并促进循环。

蒸汽可进入煮糖罐底部，在真空条件下，蒸汽泡沫上升时会膨胀，并以此帮助循环。对煮糖罐形状的影响已经进行了广泛研究，并且根据这些研究对煮糖罐的设计进行了明确改进。

糖厂在帮助我们认知煮糖方面做出了很多努力，在中央管道中设计了一个缓慢旋转的螺旋，增强了循环，方向与原始的自然循环完全相同，即通过外围向上和中心向下。这个装置已经被合并于某些真空罐中。虽然许多人相信机械循环的好处在获得第二段和第三段晶体时更为明显，在高纯度的炼糖厂糖浆中使用机械循环器也取得了非常显著的结果。在后一种情况下，尽管成粒速度的提高并不是很好，但晶体的形成和颜色要好得多。

卸料：

当结晶进程达到了实际上可以认为结晶完成的程度时，煮糖罐内的物料被排出，或者称之为“取样”。晶体和母液的混合物称为“糖膏”或“满糖”。这个糖膏或满糖进入离心机，母液会被去除离开晶体。

多段煮糖：

被离心机去除的母液在甘蔗糖厂中称为“A 蜜”，在甜菜糖厂中称为“甲原蜜”。这个液体仍然含有很多可结晶的糖，可以通过进一步的煮沸来回收。一些工厂使用两段煮糖系统，但三段煮糖系统更常见。在三段煮糖系统中，来自第一段或“A”糖的母液在第二段真空罐中煮沸结晶，称为中间煮糖罐，其中产生了第二段或“B”糖。来自 B 晶体的母液再次煮沸，产生第三段或“C”糖。C 糖的母液构成最终糖蜜。

获取“C”糖：

在这个过程中，非糖类物质在溶液中的浓度增加，增加了结晶的难度。此困难在第三段煮糖中特别明显，在许多工厂中这是回收糖的最后机会，因此在这个阶段进行了许多研究。

一般第三段或“C”糖的煮炼包括两个步骤。第一步的进行方式在不同的工厂中不同，但一种令人满意的方式是煮沸相对较少的高纯度糖浆以形成颗粒。在颗粒形成后，从第二段或“B”糖膏中提取的母液被引入煮糖罐中，并继续煮炼，直到它过饱和。过饱和的糖膏随后被排入一个助晶箱中，晶体在其中继续完全结晶。

Section 3 Crystallizer

The lower the purity of the mother liquor the slower is the rate of crystallization. It is usual to drop the masseuite into crystallizers, where additional crystallization takes place at **gradually**[1] reducing temperatures and with increased time. In the case of a well-run factor using a four-boiling system, the first massecuite might remain in the crystallizer for 6-8 h, the second for 12-16 h, the third for 30-36 h, and fourth for may be 72 h before being sent to the mixers prior to **centrifugation**[2] to remove the adhering mother liquor. (The mixer serve as a feed box for the centrifugal, it is necessary to keep the mass in motion to prevent the crystals from settling out). The syrup running off the final strike in a boiling process after it has been in the crystallizer is the by-product of the factory, and is known as blackstrap molasses. It is very heavy, possibly in the neighborhood of 90 °Bx or even more, and with an apparent purity of about 30% (sugar percent in total solids determined as Bx).

音频：Unit 6 Section 3

New words and expressions

1.	gradually	/ˈɡrædʒuəli/	*adv.* 逐步地
2.	centrifugation	/sentrɪfjʊˈɡeɪʃən/	*n.* 离心分离

Exercises

Task 1: Crosswords puzzle

请根据中文提示把下面字谜中的单词填写出来。

Across（横排）

4. 糖膏
5. 纯度，纯净
6. 逐渐地，逐步地

Down（竖排）

1. 结晶
2. 离心分离
3. 温度，气温

Task 2: One-choice question

根据对话内容，选择下列问题的正确答案。

1. The ________ of the solution determines its clarity.

   A. purity　　B. crystallization

   C. temperature　　D. mass

2. The process of forming crystals is called ________.

   A. purity　　B. crystallization

   C. temperature　　D. mass

3. The ________ of the solution can affect the rate of crystallization.

   A. purity　　B. crystallization

   C. temperature　　D. mass

Task 3: Blank filling

根据对话内容，把下列句子补充完整，每空填写一个单词。

1. The ________ takes place at gradually reducing temperatures.
2. The ________ of the solution can be determined through a purity analysis.
3. After centrifugation, the crystals may have ________ mother liquor.

Task 4: Reading comprehension

根据文章内容，回答下列问题。

1.According to the text, how long dose the second massecuite remain in the crystallizer?

____

____

____

____

____

2.What is blackstrap molasses?

____

____

____

____

____

____

Key (Unit 6 Section 3)

Task 1:

								1c								
								r								
			2c					y								
			e					s								
3t			n					t								
e			t				4m	a	s	s	e	c	u	i	t	e
m			r					l								
5p	u	r	i	t	y			l								
e			f					i								
r			u					z								
a			6g	r	a	d	u	a	l	l	y					
t			a					t								
u			t					i								
r			i					o								
e			o					n								
			n													

Task 2:

1. A 2. B 3. C

Task 3:

1. crystallization 2. purity 3. adhering

Task 4:

1. For 6-8 h.
2. The syrup running off the final strike in a boiling cycle after it has been in the crystallizer is the by-product of the factory, and is known as blackstrap molasses.

译文

第三节：助晶机

母液的纯度越低，结晶速度越慢。通常将糖膏倒入助晶机中，在逐渐降低温度和增加时间的情况下进行附加结晶。在运行良好的四段煮糖系统的工厂中，第一段糖膏可能在助晶机中停留 6 ~ 8 小时，第二段糖膏停留 12 ~ 16 小时，第三段糖膏停留 30 ~ 36 小时，第四段糖膏可能要停留72小时之后才被发送到搅拌助晶机进行离心，以去除附着的母液（搅拌助晶机作为离心机的给料罐，保持物料运动以防止晶体沉淀是必要的）。结晶周期的最后一次煮糖过程中产生的糖蜜是工厂的副产品，被称为黑糖蜜。它非常浓稠，可能达到 90 °Bx 或更高，具有约 30% 的纯度（以 Bx 为单位确定的总可溶性固体中的糖含量百分比）。

Section 4　Crystallizers

There are various types of crystallizers for these operations. They are frequently horizontal, cylindrical, or U-shaped tanks equipped with **stirring**[1] **paddles**[2]. The mixture is permitted to cool to take advantage of the lower solubility of sugar at lower temperatures. In order to **hasten**[3] the cooling. water cooling coils are frequently installed in the crystallizer. The increase in supersaturation due to cooling must not exceed the rate at which the sugar will deposit on existing crystals. Therefore, the cooling must not be too rapid since this would lead to excessive supersaturation with consequent formation of false grain. The cooling increases the viscosity of the mother liquor, and retards the rate at which the supersaturated sugar in the solution can reach the surface of the crystal. When the temperature has fallen to about 30 ℃ . Further crystallization is so slow as to be **impractical**[4].

This mass is generally too viscous to be readily centrifuged, and various methods are employed to reduce the viscosity. One of these methods is to mingle the semi-liquid massecuite with a small amount of water. Another method reduces the viscosity by elevating the temperature to about 45 ℃ . The sugar dissolves rather slowly under either of these conditions, and therefore, only a small amount of sugar is lost by being redissolved. The heating method seems to be more generally preferred since the addition of water in difficult to control, and unless properly added can cause a condition of local undersaturation and dissolve considerable amount of sugar. Proper and exact temperature is most important when massecuites are heated. Heating in the crystallizers by means of **water jackets**[5] **stationary**[6] or **rotating**[7] heating surfaces while generally employed, has certain **objectionable**[8] **features**[9]. Relatively recent process development provides rotating heating surfaces in the mixers supplying the centrifuges by which temperature adjustments are made under complete control and immediately prior to the centrifugal separation between crystals and mother-syrup.

In addition to the crystallizers already mentioned there are various other types, some of which combine the principle of a vacuum pan and a crystallizer. One recent type consists of a horizontal pan which rotates on **trunnions**[10]. After the massecuite is discharged from the pan into this type of crystallizer, the evaporation is continued until a still higher degree of supersaturation is reached. Then the steam is **shut off**[11], and cooling water is admitted to the coils. The cooling water at first is only slightly cooler than the massecuite, and this temperature is lowered gradually as the cooling progresses. In this crystallizer the final **warming**[12] of the massecuite for centrifuging can be accomplished by allowing warm water to flow through the cooling coils. An instrument known as a "Saturascope" has proved of value for the control of the temperature of the water for cooling and for warming.

音频：Unit 6 Section 4

New words and expressions

1. stir	/stɜː(r)/	*v.* 搅拌，搅和
2. paddle	/ˈpæd(ə)l/	*n.* 桨，划桨
3. hasten	/ˈheɪs(ə)n/	*v.* 促进，使加快
4. impractical	/ɪmˈpræktɪk(ə)l/	*adj.* 不切实际的
5. water jacket		水套
6. stationary	/ˈsteɪʃənri/	*adj.* 静止的；不变的
7. rotate	/rəʊˈteɪt/	*v.* 旋转，转动
8. objectionable	/əbˈdʒekʃənəbl/	*adj.* 反对的；有异议的
9. feature	/ˈfiːtʃə(r)/	*n.* 特征，特点
10. trunnion	/ˈtrʌnjən/	*n.* 轴颈；耳轴
11. shut off		关掉
12. warming	/ˈwɔːmɪŋ/	*n.* 加温，变暖

Exercises

Task 1: Crosswords puzzle

请根据中文提示把下面字谜中的单词填写出来。

Across（横排）	Down（竖排）
5. 不切实际的，做不到的	1.（使）旋转
7. 搅拌，摇动	2. 特征，特点
8. 反对的，有异议的	3. 桨，划桨
9. 静止的，固定的	4. 升温，加热
10. 促进，加速	6. 轴颈，耳轴

Task 2: One-choice question

根据对话内容，选择下列问题的正确答案。

1. What are the typical shapes of crystallizers mentioned in the text?________
   A. Rectangular. B. Triangular.
   C. Horizontal, cylindrical, or U-shaped. D. Round.
2. Why are water cooling coils installed in the crystallizer?________
   A. To warm the mixture.
   B. To slow down the cooling process.
   C. To increase supersaturation.
   D. To speed up the cooling process.
3. What happens if the cooling process is too rapid?________
   A. Excessive supersaturation with false grain formation.
   B. Sugar deposition on existing crystals.
   C. Increase in the viscosity of the mother liquor.
   D. Slow dissolution of sugar in the solution.
4. Which method is generally preferred to reduce the viscosity of the mass?________
   A. Mixing with water.
   B. Elevating the temperature.
   C. Mixing with air.
   D. Reducing the sugar content.
5. What is an advantage of using the heating method over adding water to reduce viscosity?________
   A. Faster dissolution of sugar. B. Better temperature control.
   C. Lower cost. D. Easier implementation.
6. What type of heating surfaces are generally employed in crystallizers?________
   A. Water jackets. B. Stationary heating surfaces.
   C. Rotating heating surfaces. D. Steam heating surfaces.
7. What instrument is used to control the temperature of the cooling and warming water?________
   A. Thermostat. B. Saturascope.
   C. Hydrometer. D. Thermometer.

Task 3: Blank filling

根据对话内容，把下列句子补充完整，每空填写一个单词。

1. ________________ are frequently installed in the crystallizer to hasten the cooling process.
2. The cooling process increases the ________ of the mother liquor.
3. To reduce the viscosity of the mass, one method is to mix the semi-liquid massecuite with a small amount of ________.
4. Elevating the temperature to about ________ can also reduce the viscosity of the mass.
5. The proper and exact ________ is crucial when massecuites are heated.
6. Rotating heating surfaces in the mixers supplying the ________________ allow for temperature adjustments under complete control.

Task 4: Reading comprehension

根据文章内容，回答下列问题。

1. What types of crystallizers are available for crystallization operations?

____

2. According to the passage, what methods are employed to reduce the viscosity?

____

Key (Unit 6 Section 4)

Task 1:

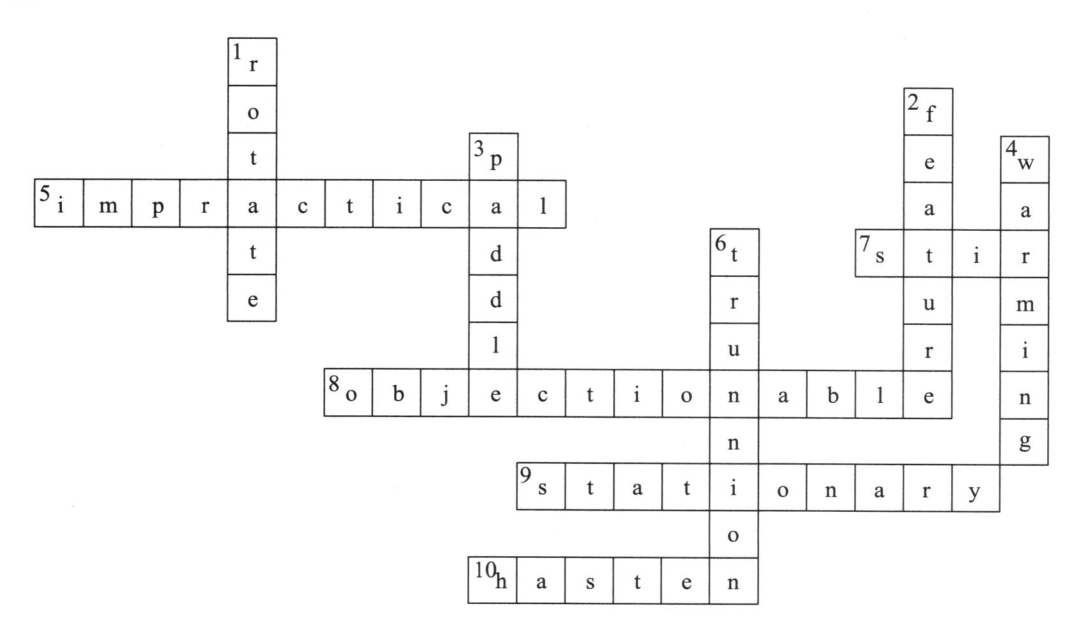

Task 2:

1. C　2. D　3. A　4. B　5. B　6. C　7.B

Task 3:

1. Water cooling coils　2. viscosity　3. water　4. 45 ℃　5. temperature　6. centrifuges

Task 4:

1. They are frequently horizontal, cylindrical, or U-shaped tanks equipped with stirring paddles.
2. One of these methods is to mingle the semi-liquid massecuite with a small amount of water. Another method reduces the viscosity by elevating the temperature to about 45 ℃ .

译文

第四节：助晶机

助晶操作通常采用各种类型的助晶机。它们通常是水平、圆柱形或U形槽样式，配备有搅拌桨。糖膏需要冷却，以利用低温下糖的溶解度变低的优势。为了加快冷却，通常在助晶机中安装水冷管。由于冷却引起的过饱和度增加不得超过蔗糖分在已有晶体上的沉积速率。因此，冷却速率不能过快，否则会导致过度过饱和形成伪晶。冷却会增加母液的黏度，降低过饱和糖溶液蔗糖到达晶体表面的速率。当温度降至约 30 ℃时，结晶的速度反而太慢而不实用。

糖膏这种物料通常太黏稠，不容易离心，因此采用各种方法降低黏度。其中一种方法是将半流质糖浆与少量水混合。另一种方法是将温度升至约 45 ℃来降低黏度。在这些条件下，糖溶解速度较慢，因此只有少量糖会被重新溶解。加热法似乎更受青睐，因为水的添加难以控制，除非适当添加，否则会导致局部不饱和状态或溶解大量糖。正确和准确的温度在加热糖浆时非常重要。虽然在助晶机中通常采用水套静止或旋转加热表面的方法，但也存在某些不良特性。比较新的工艺发展出了搅拌机上旋转加热表面供给离心机，通过这种方式可以在离心机将晶体和母液分离之前对温度进行完全控制和调整。

除了已经提到的助晶机，还有其他各种类型，其中一些结合了真空煮糖罐和助晶机的原理。最新的一种是在一个耳轴上安装旋转的煮糖罐。在将糖浆从该罐中排出到助晶机后，蒸发继续进行，直到达到更高程度的过饱和度。然后关闭蒸汽，向冷却管中注入冷却水。一开始，冷却水只比糖浆稍微凉一点，随着冷却的进行，温度逐渐降低。在这种助晶机中，允许通过温水流经冷却管来完成对糖浆的最后加热以进行离心。一种名为“过饱和计”的仪器已被证实对控制冷却和加热水的温度有重要价值。

Section 5 Identify the picture

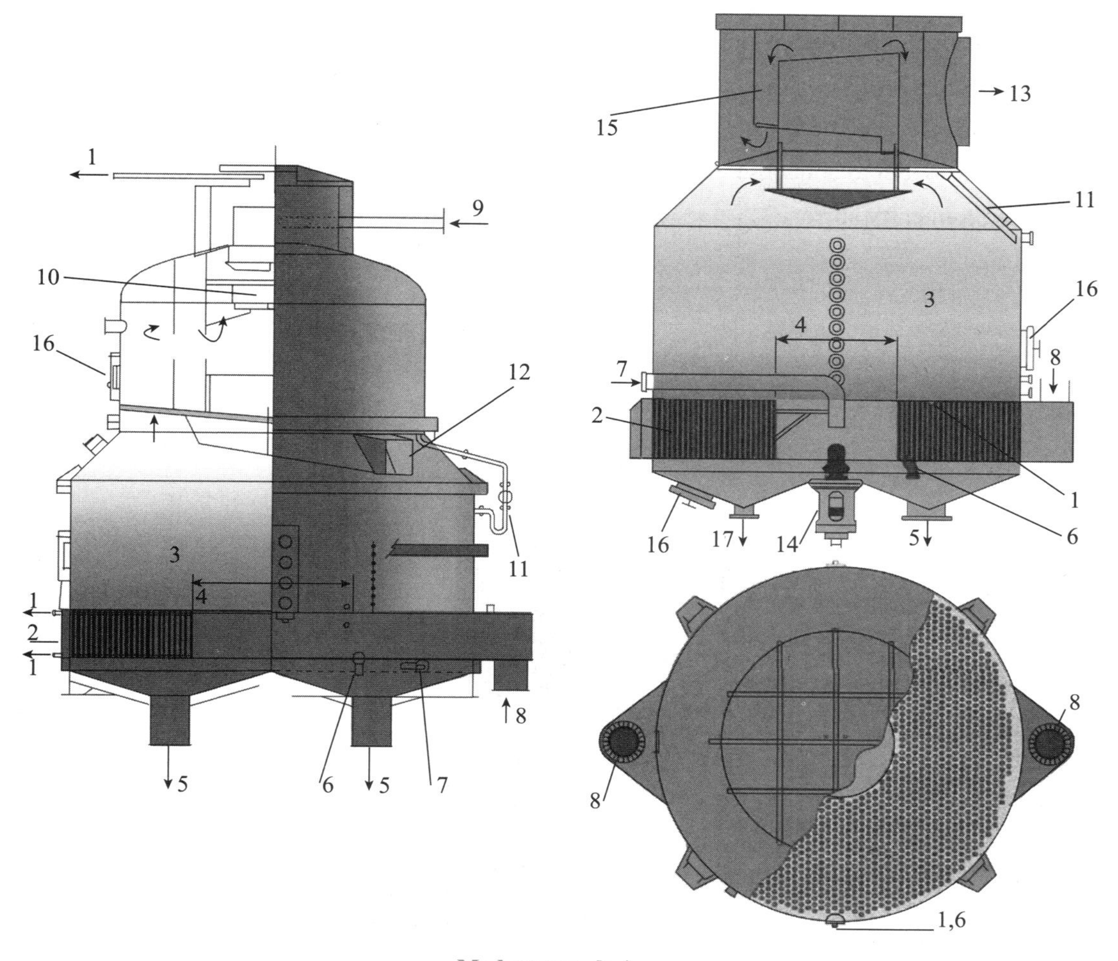

Modern pan design

New words and expressions

1. calandria /kə'lændrɪə/ *n.* 汽鼓
2. downtake /'daʊnteɪk/ *n.* 中央降液管
3. separator /'sepəreɪtər/ *n.* 分离器
4. manhole /'mænhoʊl/ *n.* 人孔
5. incondensable gas outlet 不凝气出口
6. strike level 有效容积液位
7. massecuite outlets 糖膏出口
8. condensate outlet 冷凝水出口
9. syrup feed 糖浆进料

10. steam inlet	蒸汽入口
11. condenser water inlet	冷凝水入口
12. direct contact internal condenser	直接接触式内部冷凝器
13. entrainment return line	雾沫分离器糖浆回流管
14. condenser water outlet	冷凝水出口
15. vapor outlet to condenser	蒸汽出口
16. bottom entry stirrer	内置底部搅拌器
17. cutover line	截断线

Exercises

Task1: Fill in the blanks

认真观察图片，并在空白处填上正确的单词

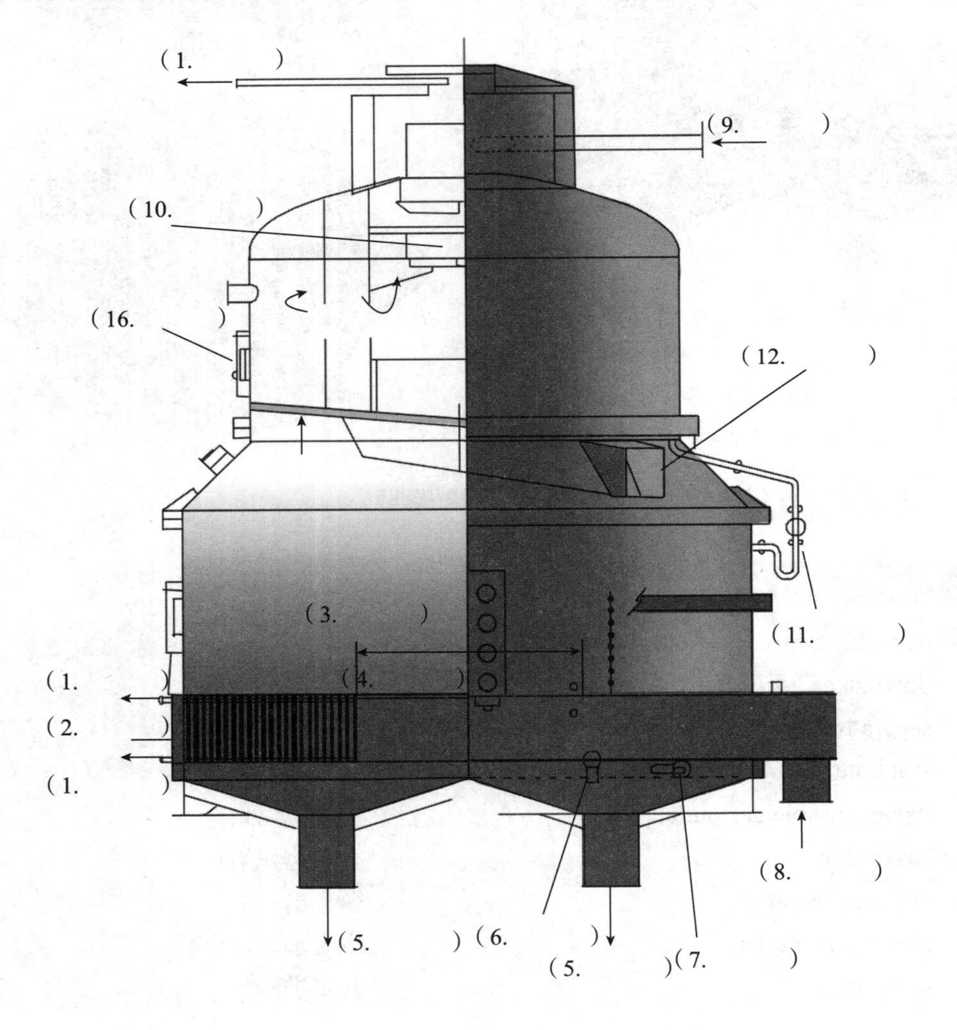

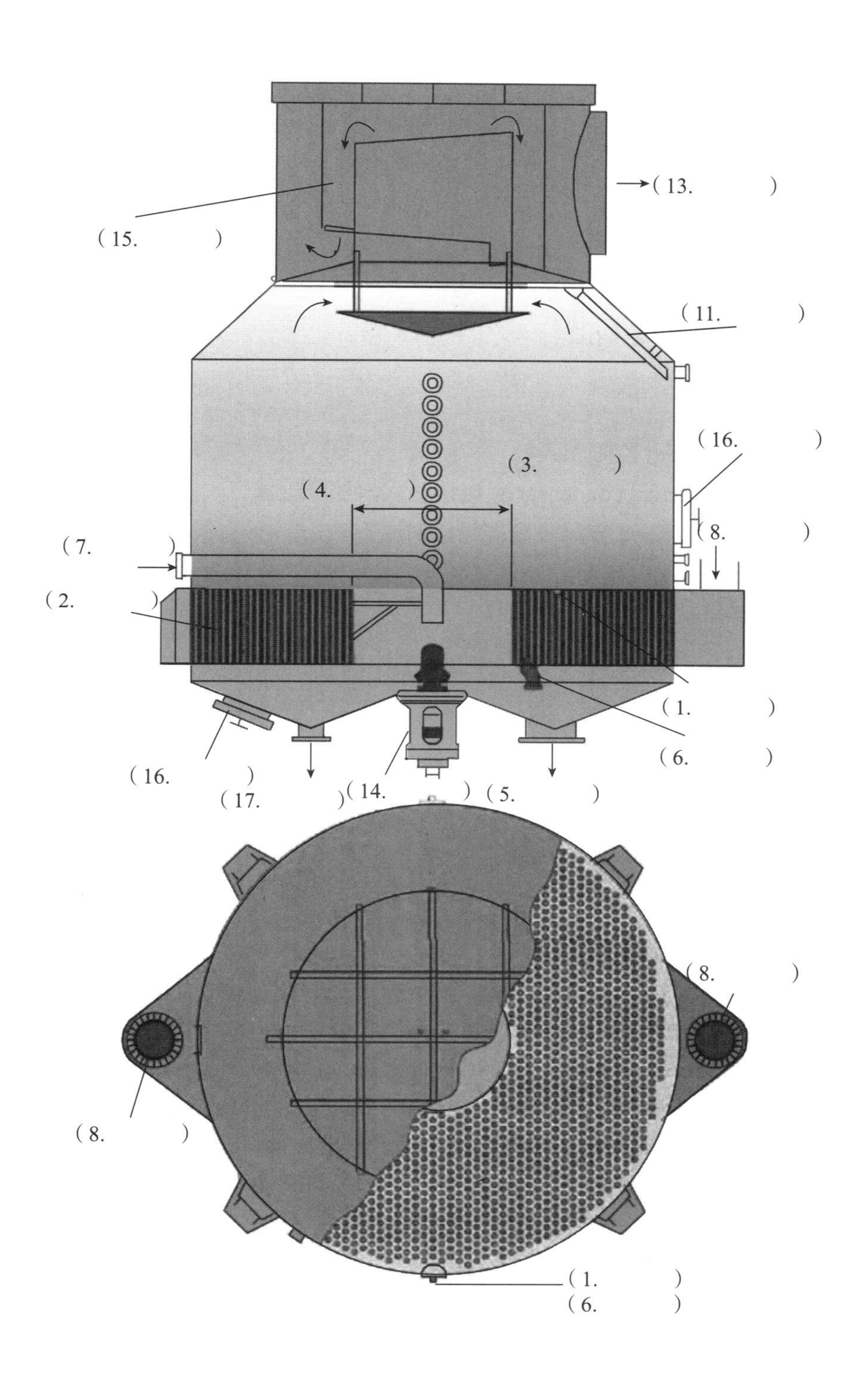
(13.　　　)
(15.　　　)
(11.　　　)
(16.　　　)
(3.　　　)
(4.　　　)
(8.　　　)
(7.　　　)
(2.　　　)
(1.　　　)
(6.　　　)
(16.　　　)
(17.　　　)
(14.　　　)
(5.　　　)
(8.　　　)
(8.　　　)
(1.　　　)
(6.　　　)

Key（Unit6　Section5）

Task1:

1. incondensable gas outlet 不凝气出口
2. calandria 汽鼓
3. strike level 有效容积液位
4. downtake 中央降液管
5. massecuite outlets 糖膏出口
6. condensate outlet 冷凝水出口
7. syrup feed 糖浆进料
8. steam inlet 蒸汽入口
9. condenser water inlet 冷凝水入口
10. direct contact internal condenser 直接接触式内部冷凝器
11. entrainment return line 雾沫分离器糖浆回流管
12. condenser water outlet 冷凝水出口
13. vapor outlet to condenser 蒸汽出口
14. bottom entry stirrer 内置底部搅拌器
15. separator 分离器
16. manhole 人孔
17. cutover line 截断线

Culture Background

The Origin of May Day Sugar Boiling Method

In 1951, renowned sugar expert Chen Shizhi worked as a technician at Dongguan Sugar Mill. Due to the lag in sugar boiling operation based on experience, he and his colleagues studied scientific methods for sugar boiling, using temperature and vacuum to control the relationship between supersaturation and crystal growth, and tried it out for powder injection and crystallization. Because it is May Day, the operation of adding powder and crystallizing is named "May Day Sugar Boiling Method". After that Chen Shizhi was transferred from the company to work at the research institute, he remained committed to this research topic. The project was jointly initiated by Guangdong Sugar Industry Company, the Sugarcane Sugar Industry Research Institute of the Ministry of Light Industry, and Dongguan Sugar Factory, with Chen Shizhi as the team leader. In 1952, the "May Day Sugar Boiling Method", characterized by full crystal nucleation and crystallization, was successfully studied, ending the outdated situation of relying solely on sensory experience for crystallization.

In 1954, an on-site meeting was held to promote the milling season. In 1955, at the first national sugar industry technology exchange conference in Beijing, Chen Shizhi introduced and read a paper on this technological achievement. Afterwards, the "May Day Sugar Boiling Method" was quickly promoted and applied in various large and medium-sized sugar factories across the country. Until now, years of practice have proven that this advanced technology has played an important role in the planned production of boiling operations, reducing the waste of molasses, and improving the production capacity of the sugar boiling process. Due to Chen Shizhi's outstanding achievements, he was named as the exemplary individual of Guangdong Sugar Industry Company in the 1955—1956 milling season. In 1956, the Ministry of Food Industry announced that the "May Day Sugar Boiling Method" was an advanced operation method of the Ministry and was promoted nationwide. Chen Shizhi was rated as the exemplary individual of the Ministry of Food Industry.

五一煮糖法的来历

1951 年，著名制糖专家陈世治在东莞糖厂当技术员，鉴于凭经验煮糖操作十分滞后，他与同事一起研究科学煮糖方法，运用温度和真空来控制过饱和度与晶体生长的关系，并尝试投粉起晶。因为当天为五一劳动节，因此，把投粉起晶操作命名为“五一煮糖法”。之后陈世治从公司转到研究所工作，仍致力于这个课题研究。由广东省糖业公司、轻工业部甘蔗糖业研究所和东莞糖厂等单位共同立项，陈世治任组长。1952 年研究成功以全晶核起晶为主要特点的“五一煮糖法”，成功研制以全结晶成核和结晶为特征的方法，结束了全凭感观经验进行起晶的落后状况。

1954 年，开展促进榨汁会议，1955 年在北京全国第一次糖业技术交流会上，陈世治把此项技术成果进行了介绍并宣读了论文。之后，“五一煮糖法”迅速在全国各大中型糖厂中得到推广应用。直到如今，多年实践证明，此项先进技术对煮炼作业计划生产、降低蜜糖损失、提高煮糖工序的生产能力起了重要作用。由于陈世治的工作成绩突出，于 1955—1956 年榨季被评为广东省制糖工业公司的先进工作者。1956 年食品工业部公布“五一煮糖法”为先进操作法在全国推广，陈世治评为食品工业部先进工作者。

Unit 7　Centrifuging

Objectives

After learning this unit, you will be able to

1. Expand your vocabulary about centrifuging and know the working process of two types of centrifuges;
2. Summarize the working process of the centrifuging.

Section　Centrifuging

After **aging**[1] in the crystallizer, the massecuite is "dried" or centrifuged in centrifuges to separate the crystals from the mother liquor. These centrifuges may be either batch ones or continuous ones.

The batch centrifuges are of various sizes and run at various speeds. Usually a batch centrifugal has a diameter（直径）of 48 in and a depth（深度）of 4 in and operates at 1 200–1 600 r/min. The thickness of massecuite against the wall of the basket（离心机筐壁） at the start of the cycle is 7 in. But after syrup has been centrifuged off, only the solid crystals remain and the thickness is reduced. The baskets of larger diameter turn slower, those of smaller diameter turn faster, in order to give approximately（大致）equal centrifugal force.

Most continuous centrifuges(连续式离心机）are essentially vertical(垂直的)cones(圆锥体) of strong screen（筛网）with the apex（尖顶） at the bottom, and is thrown out against the screen by centrifugal force（离心力）. The massecuite then progresses upward in a thin **layer**[2], hopefully only one crystal thick, so that as the diameter of the basket increases, the centrifugal force increases also. The syrup or molasses passes through the screen, and as the **purged**[3] sugar crystals get to the top of the cone, they are thrown off to be collected below the centrifuge.

In the centrifuging process, water is usually added to assist in the removal of the mother liquor adhering to（附着）the crystals. Steam may be used for the same purpose. This improves the rate of purging and quality of sugar, but some sugar is dissolved and the molasses purity **decline**[4] it is too high, for ideally there should be no sugar in the final molasse.

For the first massecuite, possibly water is scarcely needed, but with each successive lower–purity massecuite, more water must be used to wash off the ever–increasingly viscous（黏稠的）syrup. All boiled sugar should be reasonably similar, and blended to produce a fairly homogeneous raw product. In no case must so much water be used that all the molasses is rinsed off the crystal, since this thin layer of mother liquor acts as a protective barrier against **fermentation**[5]. The resultant（最终的）sugar must be dried to a point where the percentage of moisture in the whole raw sugar（原糖）is not more than 25% or at most 30% of the total non-sucrose（非蔗糖）**components**[6] of the raw sugar. (This 25% factor, known as the safety factor, was developed in Australia, and has proved to be an excellent **indicator**[7] of the keeping quality of raw sugar.)

For example, a typical analysis of raw sugar might show polarization（糖度）of 97.0 °Bx, moisture of 0.70%, ash (inorganic materials)（无机物）of 0.75%, invert (reducing sugars)（还原糖）of 0.78%, and **organic materials**[8] (determined by difference) of 0.77%. In this case the safety factor would be 0.233 and the raw would be expected to keep well. This factor is calculated thus: 0.70 /(100.0 − 97.0) = 0.233. If another raw, also polarizing（极化） 97.0 °Bx, had an analysis of 0.95% moisture, 0.68% ash,0.68% invert, and 0.69% organic materials, the safety factor would be 0.95/(100.0 − 97.0) = 0.317 which is higher, and its keeping quality would be questionable. The lower this **figure**[9] is the better. There is no **minimum**[10] safety factor, but usually a figure of 0.20 − 0.25 is obtainable with good raws（原糖）.

The last massecuite in the cycle (e.g. the third in the three–boiling system shown in Fig.7–1) is never washed, because it is desirable（令人向往的）not to dissolve any of the sucrose from the crystals and wash it into the final molasses. The sugar in the three–boiling system under consideration **is mingled with**[11] syrup to make a footing（底料）for the first two strikes（前两罐糖）, but if these strikes are to be seeded in some other way, the sugar is improved by double purging. This molasses is the principal **by–product**[12] of the raw–sugar factory, and is very viscous an black and of such purity that further crystallization of sucrose from it is economically impossible.

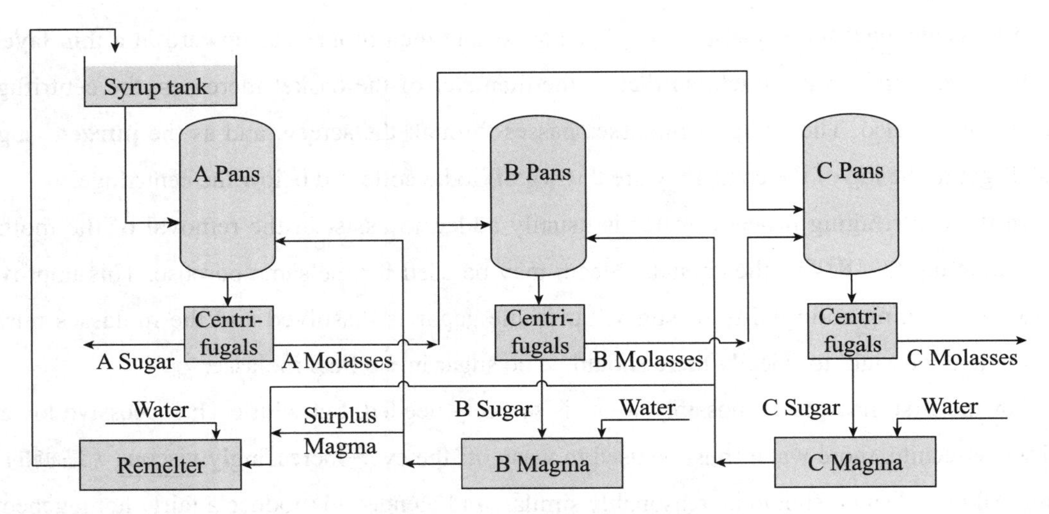

Fig 7-1 The three-boiling system

There are additional techniques used to improve centrifuging, such as hot mingling and double purging; the type used depends on the pan-boiling system employed.

Batch Versus Continuous Centrifuges:

There has been a great deal of study on the matter of application of continuous centrifuges to work formerly done by batch types. In general, it may be stated that when massecuites of very low purity are to be handled, one conventional continuous centrifuge can replace three conventional batch machines; but for massecuites of refined sugar purity, the ratio is no better than one to one, and the resultant sugar from the batch types **is superior to**[13] that from the continuous. The very great advantages of continuous centrifuges are in the area of original costs as well as in maintenance and power. In addition, they are said to be able to purge badly "smeared"（污染的，脏的） massecuites (i.e. continuing very line crystals along with the regular sized ones) which batch centrifuges cannot adequately（充分地）handle.

音频：Unit 7

New words and expressions

1. aging /ˈeɪdʒɪŋ/ *n.* 老化
2. layer /ˈleɪə/ *n.* 层
3. purge /pɜːdʒ/ *v.* 净化
4. decline /dɪˈklaɪn/ *v.* 下降
5. fermentation /ˌfɜːmenˈteɪʃ(ə)n/ *n.* 发酵
6. component /kəmˈpəʊnənt/ *n.* 成分
7. indicator /ˈɪndɪkeɪtə(r)/ *n.* 指标
8. organic materials 有机物
9. figure /ˈfɪgə(r)/ *n.* 数值
10. minimum /ˈmɪnɪməm/ *n.* 最小值

11. be mingled with	与……混合
12. by-product	副产品
13. be superior to	优于

Exercises

Task 1: Crosswords puzzle

请根据中文提示把下面字谜中的单词填写出来。

Across（横排）

6. 糖蜜
7. 最小值
9. 离心机
11. 助晶机
13. 层
14. 发酵

Down（竖排）

1. 溶解
2. 成分
3. 糖膏
4. 老化
5. 净化
8. 数值
10. 下降
12. 指标

Task 2: One-choice question

根据课文内容，选择下列问题的正确答案。

1. Batch centrifuges are of various sizes and run at various ________.
   A. diameters　B. depths　C. speeds　D. thicknesses
2. Most continuous centrifuges are essentially vertical cones made of ________ screen.
   A. weak　B. strong　C. fine　D. fragile
3. Water is usually added in the centrifuging process to assist in the removal of ________ adhering to the crystals.
   A. sugar　B. syrup　C. mother liquor　D. molasses
4. The safety factor of a raw sugar indicates its ________.
   A. purity　B. moisture content　C. keeping quality　D. sucrose content
5. The last massecuite in the cycle is not washed to avoid ________.
   A. dissolving sucrose　B. improving sugar quality
   C. further crystallization　D. excessive moisture
6. Continuous centrifuges are said to be able to purge badly "smeared" massecuites, which refers to ________.
   A. uniform-sized crystals　B. irregular crystals
   C. larger crystals　D. smaller crystals
7. Continuous centrifuges are more advantageous in terms of ________.
   A. original costs　B. sugar purity
   C. maintenance and power　D. crystal size

Task 3: Blank filling

根据课文内容，把下列句子补充完整，每空填写一个单词。

1. The massecuite is dried or ________________ in centrifugal machines to separate the crystals from the mother liquor.
2. The ____________ of massecuite against the wall of the basket at the start of the cycle is 7 in.
3. The syrup or molasses passes ________________ the screen.
4. Water is added to assist in the removal of the mother liquor ________________ to the crystals.
5. The resultant sugar must be dried to a point where the percentage of moisture in the whole raw sugar is not more than ________ %.
6. The safety factor is a good indicator of the keeping ________ of a raw sugar.
7. The last massecuite in the cycle is not washed to avoid dissolving any of the ________ from the crystals into the final molasses.
8. Continuous centrifuges are able to purge badly ________ massecuites.

Task 4: Reading comprehension

根据课文内容，回答以下问题。

Try to tell the advantages of batch centrifuges comparing with continuous centrifuges.

____

____

____

Key (Unit 7)

Task 1:

				1 d																	
2 c				i			3 m							4 a							
o				s			a							g				5 p			
6 m	o	l	a	s	s	e	s						7 m	i	n	i	m	u	m		
p				o			s				8 f			n				r			
o				l		9 c	e	n	t	r	i	f	u	g	e			g			
n				v			c				g							e			
e				e			u				u									10 d	
n							i			11 c	r	y	s	t	a	l	l	12 i	z	e	r
t							t				e							n		c	
							e											d		l	
																		i		i	
																		c		n	
																	13 l	a	y	e	r
																		t			
								14 f	e	r	m	e	n	t	a	t	i	o	n		
																		r			

Task 2:

1. C 2. B 3. C 4. C 5. A 6. B 7. C

Task 3:

1. centrifuged 2. thickness 3. through 4. adhering 5. 25 6. quality 7. sucrose

8. smeared

Task 4:

When massecuites of refined sugar purity are to be handled, batch centrifuges are more efficient, and the resultant sugar from the batch types is superior to that from the continuous.

译文

离心机分蜜

在助晶机中老化后，糖膏通过离心机的“干燥”或离心过程将晶体与母液分离。这些离心机可以是间歇式离心机，也可以是连续式离心机。

间歇式离心机大小各异，运行速度也各不相同。通常间歇式离心机的直径为 48 英寸（1 英寸 =2.54 厘米），深度为 4 英寸，运转速度为 1 200 ~ 1 600 转 / 分钟。离心机循环开始时，离心机筐壁上糖料的厚度为 7 英寸。但是，在离心分离糖浆后，只有固体晶体保留下来，厚度会减小。较大直径的筐转速较慢，较小直径的筐转速较快，以获得大致相等的离心力。

大多数连续式离心机基本上是由具有坚固筛网的垂直圆锥体构成的，通过离心机力由底部尖顶向外抛。然后，糖膏在一个薄层中向上移动，最好只有一层晶体，随着筐的直径增加，离心力也增加。糖浆或糖蜜通过筛网，当净化的糖晶体到达圆锥体顶部时，它们被抛出并在离心机下方收集。

在离心过滤过程中，通常会加水以辅助去除附着在晶体上的母液。也可以使用蒸汽来达到同样的目的。这样做可以提高净化速度和糖的质量，但是会溶解部分糖分，使糖蜜的纯度下降（即最终糖蜜中的糖分较高，理想情况下应该不含糖）。

对于第一段糖膏，可能几乎不需要加水，但是随着每段糖膏纯度的降低，必须使用更多的水来冲洗越发黏稠的糖浆。所有熬炼的糖应该是相似的，并且混合后可得到一个相当均匀的原料产品。不得使用过多的水使所有的糖蜜洗净晶体，因为这一层薄薄的母液膜起着防止发酵的保护屏障作用。最终的糖必须干燥到含水率不超过 25%，或者最多占原糖中非蔗糖成分总量的 30%（这个 25% 的因素，被称为安全因素，是在澳大利亚发展起来的，被证明是保持原糖质量的一个很好的指标）。

例如，典型原糖的分析显示，糖度为 97.0 °Bx，含水率为 0.70%，无机物（灰分）为 0.75%，还原糖为 0.78%，有机物（由差异决定）为 0.77%。在这种情况下，安全系数是 0.233，预计原糖的保存能力会很好。该系数计算如下：0.70 /（100.0 − 97.0）= 0.233。如果另一种极化为 97.0 °Bx 的原糖经分析具有 0.95% 的含水率、0.68% 的灰分、0.68% 的还原糖和 0.69% 的有机物，则安全系数将是 0.95 /（100.0 − 97.0）= 0.317，这个数值越高，其质量越具有疑问。该数值应该越低越好。没有最小的安全因素，但通常可以在 0.20 ~ 0.25 的数值中获得好的原糖。

煮炼周期中的最后一段糖膏（例如，图 7-1 所示的三段煮糖系统中的第三次煮炼）永远不会洗涤，因为不希望溶解任何蔗糖并将其洗入最终的糖蜜中。三段煮糖系统中的糖与糖浆混合，以制成前两个糖膏的底料，但如果这些糖以其他方式进行配制，通过双层净化可以改善糖的质量。这种糖蜜是原糖厂的主要副产品，非常黏稠且黑，纯度很低，因此从经济上考虑无法进一步提取蔗糖。

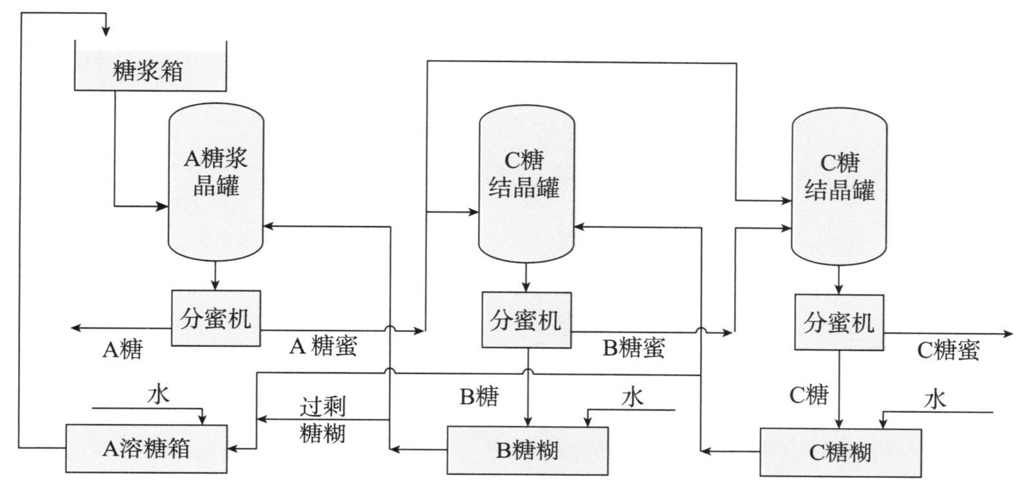

图 7-1　三段煮糖系统

还有其他用于改善离心过滤的技术，如热混合和双层净化；所使用的类型取决于所采用的煮炼系统。

间歇与连续式离心机：

有关连续式离心机替代间歇式离心机的研究已经相当多。通常情况下，当处理质量极低的糖浆时，一个传统的连续式离心机可以代替三个传统的间接机；但对于含糖纯度较高的糖浆，比例不会好于一对一，并且间歇式离心机得到的糖优于连续式离心机得到的糖。连续式离心机非常大的优势在于原始及维护和动力成本低。此外，据说它们能够净化“严重污损”的糖浆（即连续生产非常小的晶体和常规大小晶体），而间歇式离心机无法充分处理。

Culture Background

Separation Technology of Molasses and Crystallization before Industrialization

India often places massecuite in rough jute bags with heavy objects placed on them to squeeze out molasses. In China, the sugar massecuite is placed in a conical jar to discharge the molasses, which are repeatedly discharged and dried to produce refined white sugar filtered with clay.

工业化以前的结晶与糖蜜分离技术

印度多将糖膏置于粗糙的麻袋中，袋子上放置重物以挤压出糖蜜。中国是将糖膏装在圆锥形的坛子中排出糖蜜，糖蜜排出和晒干不断重复滤过黏土，以生产精炼白糖。

Unit 8 Storage

Objectives

After learning this unit, you will be able to

1. Expand your vocabulary about storage;
2. Understand the essential measures for storage;
3. Summarize the essential measures for storage.

Section 1 Storage

Storing the **raws**[1] sugar in the producing country until sufficient volume is **accumulated**[2] for shipment is usually done by **piling**[3] in dry, well-**ventilated**[4] **warehouses**[5]. This is done by means of overhead belts equipped with **diverters**[6] or **cross-belts**[7] or by means of a **slinger**[8], which **throws** the sugar **off** a very high-speed belt in a strong stream（强劲的水流）**onto**[19] a pile which may be as high as thirty feet or more. There is an **attendant**[9] advantage to such a slinger, because some cooling and drying is effected in the throwing.

The raws so stored should adhere to（遵守）the safety factor **delineated**[10]. The material should be cooled **as much as possible**[20], because warm raw sugar in a pile, especially if at all moist（潮润的）, tends to **pack into**[21] a very hard mass, and worse, to increase in color due to **caramelization**[11] at the elevated temperature（升高的温度）.

Attention must be paid to the possibility of fire. Raws stored in bags should be given the same care as **bulk**[12] piles; the **susceptibility**[13] to fire is very much greater; the juice and other containers support combustion（燃烧）, and the **voids**[14] storage between the units provide a free flow of air for combustion. In a fire in bulk raw, if caught reasonably soon, damage caused by the **sprinklers**[15] is usually light, because the surface sugar packs, permitting the water to flow off. Similar sprinkler damage in a pile of straw（稻草）in hags results in a sad **mess**[16], because the water flow down between the bags **permeating**[17] the entire **stack**[18] to the ground, and serious fermentation results if the raws are not refined quite promptly.

音频：Unit 8 Section 1

New words and expressions

1. raw	/rɔːs/	*n.* 原始的
2. accumulate	/əˈkjuːmjəleɪt/	*v.* 积累，堆积
3. pile	/paɪl/	*v.* 堆放
4. ventilate	/ˈventɪleɪt/	*v.* 通风
5. warehouse	/ˈweəhaʊs/	*n.* 仓库
6. diverter	/daɪˈvɜːtə(r)/	*n.* 转移器
7. cross-belt		十字交叉运输带
8. slinger	/ˈslɪŋə(r)/	*n.* 抛投机
9. attendant	/əˈtendənt/	*adj.* 伴随的，随之产生的
10. delineate	/dɪˈlɪnieɪt/	*v.*（详细地）描述；标注
11. caramelization	/ˌkærəməlaɪˈzeɪʃən/	*n.* 焦化；产生焦糖
12. bulk	/bʌlk/	*adj.* 大批的，大宗的
13. susceptibility	/səˌseptəˈbɪləti/	易感性
14. void	/vɔɪd/	*n.* 空间
15. sprinkler	/ˈsprɪŋklə(r)/	*n.* 喷头
16. mess	/mes/	*n.* 混乱
17. permeate	/ˈpɜːmieɪt/	*v.* 渗透
18. stack	/stæk/	*v.* 堆积
19. throw...off...onto		把……从……上扔到……上
20. as much as possible		尽可能地
21. pack into		压紧，将物品紧密地放入某个容器中

Exercises

Task 1: Crosswords puzzle

请根据中文提示把下面字谜中的单词填写出来。

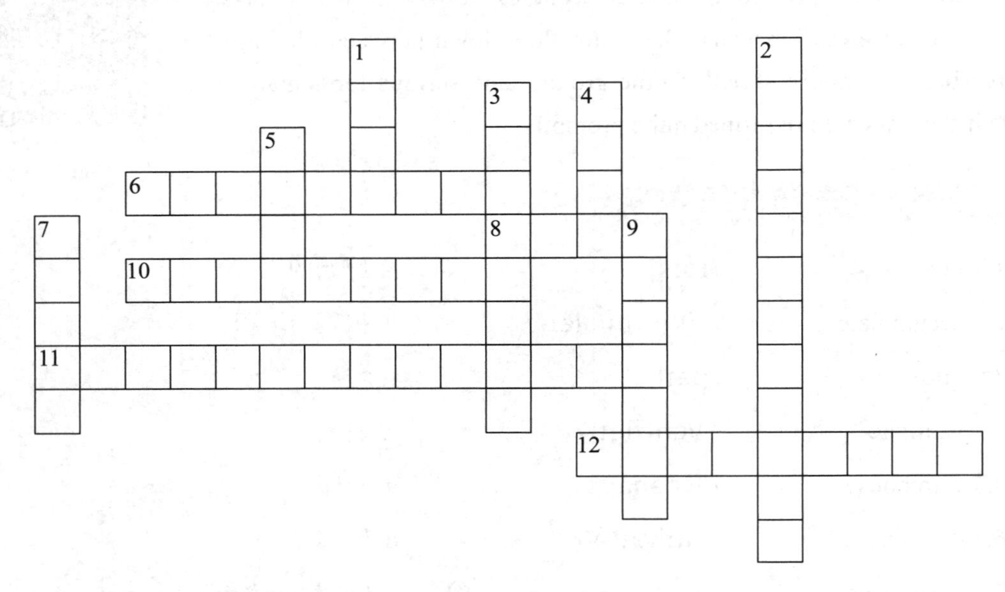

Across（横排）

6. 喷头
8. 混乱
10. 仓库
11. 焦化
12. 通风

Down（竖排）

1. 大批的
2. 发酵
3. 渗透
4. 原糖
5. 堆放
7. 堆积
9. 抛投机

Task 2: One-choice question

根据对话内容，选择下列问题的正确答案。

1. The primary purpose of storing raw materials before shipment is to________.
   A. increase the color of the raw sugar
   B. decrease the temperature of the raw sugar
   C. accumulate a sufficient volume for shipment
   D. prevent fermentation of the raw sugar
2. Which of the following is NOT a method of storing raw materials?________
   A. Piling them in dry, well-ventilated warehouses.
   B. Using overhead belts with diverters.

C. Using an overhead slinger.

D. Storing them in bags.

3. The word “permeating” in the last sentence means ________.

A. spreading through something

B. packing tightly

C. increasing in color

D. causing fermentation

Task 3: Blank filling

根据对话内容，把下列句子补充完整，每空填写一个单词。

1. The raw materials should adhere to the ________ factor mentioned above.
2. Warm raw sugar in a pile can ________ into a hard mass.
3. Raws stored in bags are more susceptible to ________.
4. Sprinkler damage to a pile of raws in bags can cause a ________ mess.
5. Serious fermentation can occur if the raws are not ________ quite promptly.

Task 4: Reading comprehension

根据对话内容，回答下列问题。

1. What happens to warm raw sugar in a pile if it is moist?

____

____

____

____

____

____

____

____

2. What is one advantage of using a slinger for storing raw sugar?

____

____

____

____

____

____

____

____

Key (Unit 8 Section 1)

Task 1:

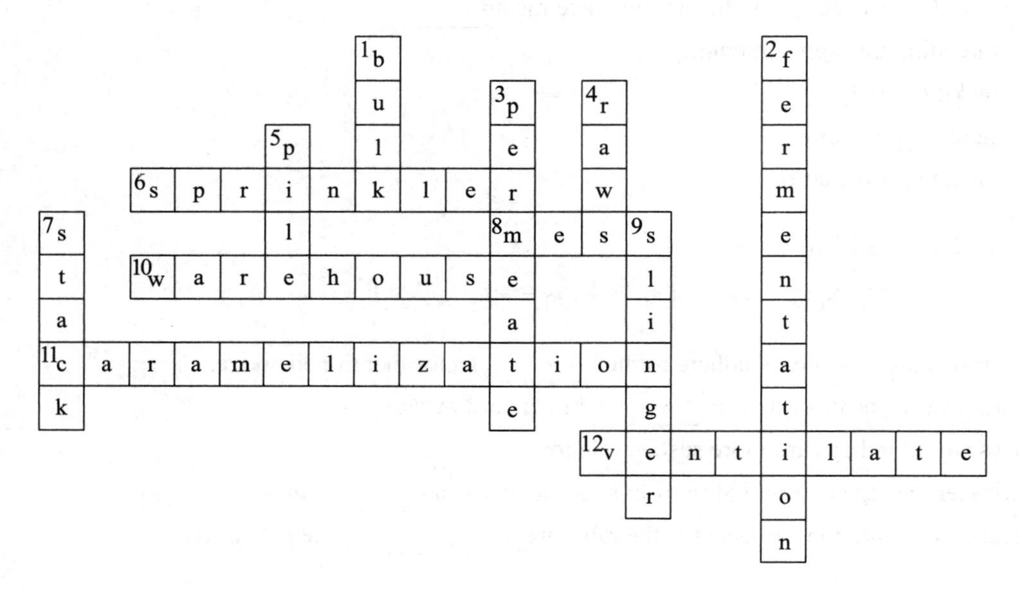

Task 2:

1. C　2. D　3. A

Task 3:

1. safety　2. pack　3. fire　4. sad　5. refined

Task 4:

1. It tends to pack into a hard mass.
2. It reduces the chances of fire.

译文

第一节：储存

通常将生产的原糖堆放在干燥、通风良好的仓库中，直到数量达到装运要求。原糖的堆放是通过配备分流器、交叉带传送带或抛投机完成的。抛投机以强劲的水流将原糖从高速皮带上抛到大约30英尺或更高的原糖堆上。使用抛投机的好处是在抛料过程中，原糖得到了一定程度的冷却和干燥。

如此储存的原糖应符合规定的安全系数。原糖应该尽可能地冷却，因为热的原糖堆在一起，在潮湿的环境下，往往会压成非常硬的团块，更糟糕的是，高温会促使原糖焦糖化，使颜色加深。

必须注意发生火灾的可能性。袋装的原糖和散装的原糖堆都需要相同的管理，袋装的原糖对火灾的易感性要大得多。糖浆和其他容器会助燃，仓库的存储单元之间的空隙为燃烧提供流动的空气。散装原糖发生火灾，如果发现及时，通常是轻微造成喷头损坏，因为表面的糖会使水流动。类似的喷头损坏导致的火灾则会在一堆稻草上引发严重的混乱，因为水会从袋子之间流下，渗透至地面，如果不及时把原糖送往生产，就会导致严重的发酵事故。

Section 2 Changes during storage

Sugar produced in factories has to be stored in warehouses for months together during which periods **atmospheric**[1] changes like hot summer months, **humid**[2] rainy weather and cold winter are inevitable. The crystal sugar of high quality must retain the **lustrous**[3] appearance and **whiteness**[4] during storage until it reaches the consumers but the white sugar produced by carbonation factories has undergone **deterioration**[5] in colour during storage. Compared to the juice **sulphitation**[6] process the double carbonation（双碳酸法）offers distinct advantage is respect of consistently producing good quality white sugar under **varying**[7] conditions of juice composition, but as far as keeping quality of white sugar is concerned the carbonation sugar under-goes faster deterioration of colour than the sulphitation sugar which retains whiteness and lustre for a fairly long period. In the **conventional**[8] double carbonation, the juice after second carbonation at about pH 8.0 - 8.5 is **neutralised**[9] with SO_2, showed that the reduction of SO_2, in stored sugar is accompanied by **development**[10] of brownish ting（淡褐色的） particularly in hot periods. This development of colour to be due to slow caramel formation resulting from carbonates in sugar reacting with reducing sugars. Carbonation sugar is thus more **susceptible**[11] to caramel formation due to presence of more **alkaline**[12] carbonates as compared to sulphitation sugar since the carbonate content of the former is 8 to 16 times that in the latter. The carbonates of potash and sodium（钾和钠）embedded in sugar crystals catalyse（催化）the carmelization（焦化反应）of reducing sugar.

Caking of sugar（糖结块）in bags is a phenomenon associated with（与……相关联）external factors like weather conditions as also the nature of sugar crystals. Sugar, due to its **hygroscopic**[13] character is exposed to atmospheric conditions of high relative **humidity**[14] it will absorb moisture. Similarly, the sugar crystal part with the moisture whenever they come in contact with atmosphere of low relative humidity. In consonance with this, whenever packed sugar which contains originally high moisture or has absorbed moisture after packing is exposed to atmosphere of low relative humidity, the syrup surrounding the sugar crystals parts with some of moisture and reaches zone of high **supersaturation**[15]. The sugar crystals being closely packed, crystallisation in the syrup results in joining of sugar crystals, eventually leading to cake formation. The blocks of sugar cakes are difficult to dislodge from the packing bags and present problems in marketing. The factors favoring the cake formation of sugar crystals are: ① small grain size; ② **impurities**[16] in sugar crystals like reducing sugars which contribute to the hygroscopic character of sugar; ③ higher initial moisture of sugar at the time of packing than specified (> 0. 04%); ④ higher relative humidity of atmosphere followed by dry weather conditions of low R. H.

In the absence of any **remedial**[17] measures in the case of caking of sugar following preventive measures are essential: ① Sugar should be dried well which means efficient removal of unbound moisture. Prior to packing, sugar must be cooled to 38–40 ℃, since hot sugar bagging leads to cake formation. ② In view of (鉴于) the role of impurities like reducing sugars (还原糖) in increasing hygroscopicity of white sugar, in the process every effort is essential to minimize the impurities in sugar crystals. ③ Small grain size favors moisture absorption and in case the grain size is to be maintained small to suit market needs. ④ Irrespective of the size of sugar grain, the proper grading of sugar has to be taken care of to avoid mixing of different size grains.

音频：Unit 8　Section 2

New words and expressions

1. atmospheric	/ˌætməˈsfɪrɪk/	*adj.* 大气的
2. humid	/ˈhjuːmɪd/	*adj.* 潮湿的
3. lustrous	/ˈlʌstrəs/	*adj.* 有光泽的
4. whiteness	/ˈwaɪtnəs/	*n.* 白度
5. deterioration	/dɪˌtɪriəˈreɪʃən/	*n.* 恶化
6. sulphitation	/ˌsʌlfɪˈteɪʃən/	*n.* 硫化
7. varying	/ˈveəriɪŋ/	*adj.* 不同的
8. conventional	/kənˈvenʃənəl/	*adj.* 传统的
9. neutralise	/ˈnjuːtrəaɪz/	*v.* 中和
10. development	/dɪˈveləpmənt/	*n.* 发展
11. susceptible	/səˈseptəbəl/	*adj.* 易受影响的
12. alkaline	/ˈælkəlaɪn/	*adj.* 碱性的
13. hygroscopic	/ˌhaɪgrəˈskɒpɪk/	*adj.* 吸湿性的
14. humidity	/hjuːˈmɪdəti/	*n.* 湿度
15. supersaturation	/ˌsupərˌsætʃəˈreɪʃən/	*n.* 超饱和度
16. impurity	/ɪmˈpjʊərɪti/	*n.* 杂质
17. remedial	/rɪˈmiːdiəl/	*adj.* 补救的

Exercises

Task 1: Crosswords puzzle

请根据中文提示把下面字谜中的单词填写出来。

Across（横排）	**Down（竖排）**
2. 杂质	1. 超饱和度
4. 焦糖	3. 中和
8. 湿度	5. 潮湿的
10. 硫化	6. 大气的
11. 碱性的	7. 白度
	9. 不同的

Task 2: One-choice question

根据文章内容，选择下列问题的正确答案。

1. What are the atmospheric changes mentioned in the passage?________

A. Hot summer months and cold winter.

B. Humid rainy weather and hot summer months.

C. Cold winter and humid rainy weather.

D. Hot summer months, humid rainy weather and cold winter.

2. What happens to the white sugar produced by carbonation factories during storage?________

A. It retains its whiteness and lustre.

B. It undergoes faster deterioration in colour.

C. It becomes lustrous and transparent.

D. It becomes brownish in colour.

3. What is the advantage of double carbonation in producing white sugar?________

A. Consistently producing good quality white sugar.

B. Producing sugar with a longer shelf life.

C. Retaining the whiteness and lustre of sugar.

D. Avoiding deterioration in colour during storage.

4. What is the cause of the development of brownish tinge in stored sugar during hot periods?________

A. Slow caramel formation.

B. Reduction of SO_2.

C. Presence of alkaline carbonates.

D. Reaction with reducing sugars.

5. Why is carbonation sugar more susceptible to caramel formation compared to sulphitation sugar?________

A. It has a higher initial moisture content.

B. It contains more alkaline carbonates.

C. It undergoes a double carbonation process.

D. It is stored for a longer period.

6. What is the phenomenon associated with the caking of sugar in bags?________

A. Exposure to weather conditions.

B. Moisture absorption of sugar crystals.

C. Joining of sugar crystals in the syrup.

D. Difficulties in marketing.

Task 3: Blank filling

根据文章内容，把下列句子补充完整，每空填写一个单词。

1. The ________ conditions can affect the quality of stored sugar.

2. The white sugar produced by carbonation factories undergoes ________ in colour during storage.

3. The double carbonation process offers the advantage of consistently producing good quality white sugar ________ varying conditions.

4. Carbonation sugar undergoes faster ________ of colour than sulphitation sugar.
5. The development of colour in stored sugar is due to slow ________ formation resulting from carbonates in sugar reacting with reducing sugars.
6. Caking of sugar is caused by exposure to atmospheric ________ and the nature of sugar crystals.
7. The factors that favor the cake formation of sugar crystals include small ________ size and higher impurities in sugar crystals.
8. In order to prevent the caking of sugar, it is essential to dry the sugar well and remove ________ moisture.

Task 4: Reading comprehension

根据对话内容，回答下列问题。

1. What are the essential preventive measures for the caking of sugar?

____

2. What are the factors that favor the cake formation of sugar crystals?

____

Key（Unit 8　Section 2）

Task 1:

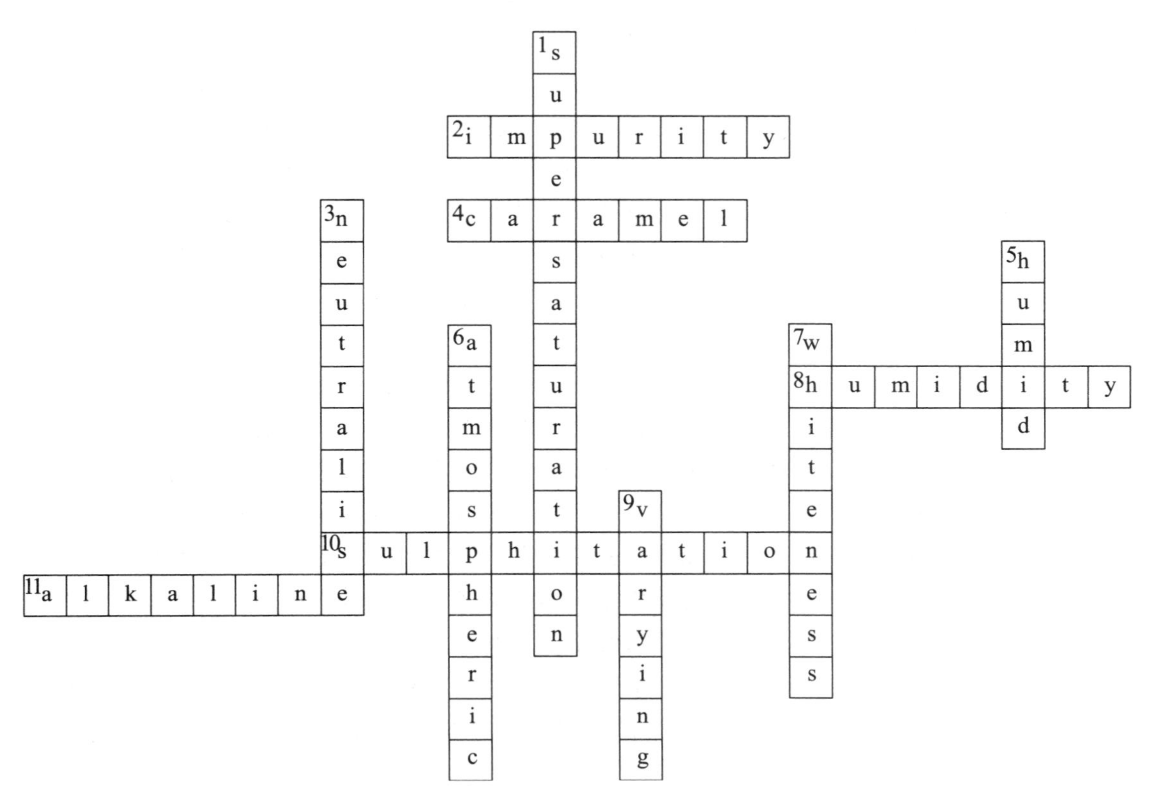

Task 2:

1. B 2. B 3. A 4. A 5. B 6. C

Task 3:

1. atmospheric 2. deterioration 3. under 4. deterioration 5. caramel 6. conditions 7. grain 8. unbound

Task 4:

1. ① Drying the sugar and cooling it before packing
   ② Decreasing the impurities in sugar crystals
   ③ Maintaining small grain size of sugar
   ④ Mixing different size grains of sugar
2. Small grain size and higher initial moisture

译文

第二节：储存期间的变化

工厂生产的糖必须在仓库中存放数月，而在此期间不可避免地会发生如炎热的夏季、潮湿的雨季和寒冷的冬季等大气变化。高质量的结晶糖必须在存储期间保持光泽和白度，直到到达消费者手中。然而，碳法厂生产的白糖在存储期间颜色会发生恶化。与亚硫酸法工艺相比，双重碳酸法在蔗汁成分变化条件下始终能产生优质白糖，但就白糖保持品质而言，硫化工艺下的白糖相对碳酸化工艺下的白糖有更长的保持白色和光泽的时间。在常规的双碳酸化过程中，第二次碳化后的蔗汁在pH值为8.0～8.5时，通过二氧化硫中和，存储糖中二氧化硫的减少特别是在炎热时期会出现淡褐色的现象。这种颜色的变化是由糖中碳酸盐与还原糖发生缓慢的焦糖化反应所致。相比之下，碳法糖更容易发生焦糖化，因为碳酸化糖中的碱性碳酸盐含量比硫化糖多8至16倍。糖晶体中的钾和钠碳酸盐催化还原糖的焦糖化反应。

糖在袋子中结块是与天气条件以及糖晶体性质等因素相关的现象。由于糖具有吸湿性，暴露在相对湿度较高的大气条件下，它会吸收水分。同样地，只要糖晶体接触到相对湿度较低的空气，它们就会释放出水分。根据这一点，包装的糖含有最初的高湿度或者在包装后吸收了水分，当暴露在相对湿度较低的大气中时，包裹在糖晶体周围的糖浆会释放一些水分，达到高过饱和度。由于糖晶体紧密堆积，糖浆中的结晶会导致糖晶体的连接，最终形成结块。结块的糖块难以从包装袋中脱落，给市场销售带来问题。影响糖晶体结块的因素有：①较小粒子；②糖晶体中较高的杂质，如还原糖增加了糖的吸湿性；③包装时糖的初始含水量高于规定值（>0.04%）；④低相对湿度的干燥天气后出现大气相对湿度较高。

在没有任何补救措施的情况下，防止蔗糖结块的措施十分必要：①必须使糖充分干燥，即有效去除未结合的水分。在包装前，糖必须冷却到 38 ～ 40℃，因为热糖装袋会导致结块。②鉴于还原糖等杂质在提高白糖吸湿性方面的作用，必须在过程中努力减少糖晶体中的杂质。③晶粒较小有利于吸湿，但晶粒大小要保持小以适应市场需求。④无论糖颗粒的大小如何，都必须注意糖的正确分级，以避免不同大小颗粒的混合。

Culture Background

Effect of Ancient Chinese Sugar-making Technology to Ancient Ryukyu and Japan

In the history of East Asian agriculture, China often spread the technology of planting economic crops and processing products to neighboring countries through cultural, economic and trade means, effectively promoting the agricultural development of neighboring countries, with the dissemination of sugar making method being one of the most typical cases. For example, the ancient Ryukyu Kingdom and Japan gained the most advanced sugar making technology in their interactions with China, which had a profound impact on the historical development of these two countries.

古代中国制糖技术对古代琉球和日本的辐射

在东亚农业史上，中国经常将经济作物种植及制品加工技术通过文化、经贸等方式传播到周边国家，有力推动了周边国家的农业发展，其中制糖法的传播就是最为典型的案例之一。比如，古代琉球王国、日本在与中国的交往中，获取到当时最为先进的制糖技术，对这两个国家的历史发展影响颇为深远。

参考文献

［1］陆冬梅 . 制糖工程专业英语［M］. 北京：中国轻工业出版社，2012.
［2］孙卫东 . 制糖专业英语（广西大学校本教材）［M］. 南宁：广西大学，2003.
［3］Peter Rein. Cane Sugar Engineering［M］.Germany Berlin：Bartens，2007.